To Margaret

CELL
BIOLOGY

JOHN W. KIMBALL

ADDISON-WESLEY
PUBLISHING COMPANY
Reading, Massachusetts
Menlo Park, California
London • Don Mills, Ontario

This book is in the
ADDISON-WESLEY SERIES IN LIFE SCIENCE

ISBN 0-201-03675-4
FGHIJKLMNO-HA-7987654

PREFACE

In recent years the teaching of introductory biology in both schools and colleges has been undergoing rapid change. The variety of innovations in content and organization has created a need for a variety of short, specialized textbooks from which the teacher can choose to meet the needs of his particular course. This book is an attempt to introduce the beginning student to the structure and function of the living cell. The material in the book is derived largely from those portions of my general text, *Biology* (Second Edition, Addison-Wesley, 1968), that concentrate on the cell itself.

The first section of this book is devoted to a study of cell structure at several levels of organization. In the belief that an understanding of the structure and functioning of cells depends, at least in part, upon an understanding of the properties of matter at lower levels of organization, the student is first introduced to the essentials of atomic structure and the formation of small molecules. This is followed by an examination of the structure of the macromolecules of living things. Then the cell organelles are studied and an effort made to relate their structure to that of the macromolecules of which they are composed.

Part II of the book is devoted to cell metabolism. The mechanisms by which materials pass to and from the cell and its environment and how these materials are transformed within the cell are studied first. Then the trapping of the sun's energy by photosynthesis and the release of this trapped energy by cellular respiration are examined in detail.

Part III is devoted to the general problem of the genetic control of cells. Chapter 7 examines the role of DNA as a repository of genetic information and the role of chromosomes in the cells of higher organisms. In Chapter 8, the way in which the information encoded in DNA is translated into the traits of the organism is studied. This is followed by a consideration of possible mechanisms by which genetic information is *selectively* translated, both in response to the changing needs of the differentiated cell and during the process of differentiation itself. The material of these two chapters has to a large extent been written especially for this book, as the related topics in the parent volume were so interwoven with organismic biology.

I am grateful to my former students and to my colleague Harper Follansbee for their thoughtful criticisms and suggestions, especially of the new material and organization in the final two chapters. I also wish to thank the staff at Addison-Wesley for their cooperation and skill in all phases of producing the book. My heartfelt thanks go also to my wife, Margaret, whose efforts, both direct and indirect, contributed so much to this project.

The Biological Laboratories   J. W. K.
Harvard University

CONTENTS

PART I

THE ORGANIZATION OF LIFE

Crystals of a salt adhering to the bottom of a one-liter flask (approximately life-size). The pattern is a reflection of the orderly stacking of positive and negative ions in the crystal lattice (see Section 1-6). (Courtesy Dr. Waldo E. Cohn and Photographic Services Dept. of Oak Ridge National Laboratory.) ▶

CHAPTER 1

THE CHEMICAL BASIS
OF LIFE: PRINCIPLES

You may think it odd that a book on the subject of the structure and function of cells should begin by examining some of the basic principles of chemistry. There was a time when virtually all our knowledge of cells came from the keen observations of biologists working with their preservatives, tissue-slicing machines, a few stains, and microscopes. However, as the analytical techniques of the chemists and physicists were turned on biological materials, it became possible to analyze cells in terms of structural units far smaller than those visible under the light microscope. These units are atoms and molecules.

As a result of such studies, two things have become clear.

1) Although living things and, to a lesser degree, the cells of which they are composed are quite diverse in their appearance, their basic chemical organization is remarkably similar.

2) Although the chemical organization of living matter is very complex, it is based upon the same materials and principles as those found in the world of nonliving things.

These two discoveries have had far-reaching consequences. They have shifted the attention of biologists from the many ways in which living things differ from one another to the many ways in which they are similar. This shift in approach has gone hand in hand with an enlargement of our way of looking at living things. No longer can a biologist be content simply to study **morphology,** that is, the way in which organisms are put together out of cells, tissues, and organs. Now he must ask how these various structural parts of the organism work; that is, he must also examine their **physiology.** To find the answers he must understand the chemical makeup of the living cell and the chemical principles underlying its activity.

1–1 ELEMENTS

All the things found on this planet—air, sea water, soil, rocks, etc.—are composed of one or more fundamental building blocks of matter, the elements. Despite the large number of kinds of rocks, minerals, etc., only 90 different elements have been found occurring naturally on earth. Actually, 103 different elements are known, but 13 of these have been man-made in the laboratories of nuclear physics.

We can define an element as a substance that cannot be further decomposed by chemical means. Iron, oxygen, aluminum, and silicon are four common elements found on or near the surface of the earth. To say that everything on this earth is composed of, and is thus reducible to, one or more of 90 different elements does not exclude living things. They, too, are made up of elements. Of the 90 present generally on earth, however, only about one-quarter are found in living organisms. Not only is the number of elements found in living organisms smaller, but the relative prevalence of these elements is different. Figure 1–1

Distribution of elements in the crust of the earth, including land, air, and water

	% by weight
Oxygen (O)	49.5
Silicon (Si)	25.3
Aluminum (Al)	7.5
Iron (Fe)	5.08
Calcium (Ca)	3.39
Sodium (Na)	2.63
Potassium (K)	2.40
Magnesium (Mg)	1.93
	97.69%
Hydrogen (H)	0.87
Titanium (Ti)	0.63
Chlorine (Cl)	0.19
Phosphorus (P)	0.12
	99.50%
Manganese (Mn)	0.090
Carbon (C)	0.080
Sulfur (S)	0.060
Barium (Ba)	0.040
Chromium (Cr)	0.038
Nitrogen (N)	0.030
Fluorine (F)	0.026
Zirconium (Zr)	0.023
Strontium (Sr)	0.020
Nickel (Ni)	0.018
Zinc (Zn)	0.017
Vanadium (V)	0.018
Copper (Cu)	0.010
Total	99.96%

Distribution of elements in living things

	% by weight
O	65
C	18
H	10
N	3
Ca	2
P	1
	99%
K, S, Cl, Na, Mg, Fe	0.9%
Mn, Cu, I (iodine), Co (cobalt), Zn; B (boron), Mo (molybdenum)	0.1%
Total	100.0%

Fig. 1-1

The distribution of elements in the crust of the earth compared with their distribution in a living organism.

compares the elemental composition of the crust of the earth with that typical of a living organism. Obviously, living organisms are not simply a reflection of the chemical composition of their environment. Some elements found abundantly in the nonliving world (e.g. aluminum) play no role in living things. On the other hand, carbon, hydrogen, and nitrogen are found greatly concentrated in living organisms. In fact, these three elements plus oxygen make up 96% of all living material. It looks as though one feature of life is its ability to accumulate and concentrate elements that are relatively rare in the nonliving environment.

Other elements are essential to life, too. Calcium, phosphorus, sodium, potassium, magnesium, chlorine, sulfur, and iron are always found in living things. Added to the previous four, these account for 99.9% of the weight of living organisms. The remaining one-tenth of one percent is made up of a few other elements such as copper, manganese, zinc, cobalt, iodine, vanadium, fluorine, and (in plants only) boron and molybdenum. Because these latter elements are present in such tiny quantities, they are called trace elements. This does not mean that they do not play an important role in life. It is simply that a little goes a long way.

1-2 ATOMS

Each of the 103 elements known to man is composed of one particular kind of atom. We can define an atom as the smallest part of an element that can enter into combinations with other elements. (More of this in a moment.) For ease in discussing elements and the atoms of which they are composed, we use symbols, one for each element (Fig. 1-2).

We have said that elements cannot be further decomposed by chemical means. The same is true, of course, of the atoms of which they are composed. Atoms can, however, be decomposed by the use of more violent, physical methods. Years of such work have shown that atoms, too, have a very definite and orderly structure.

The simplest atom, the hydrogen atom, consists of a single positively charged particle, the **proton,** surrounded by a single negatively charged particle, the **electron.** The two charges are equal in magnitude so the atom is electrically neutral. Almost all the weight of the atom is accounted for by the weight of the proton.

The element with the next most complicated atom is the gas helium (He). Its atoms have two protons and also two other particles called **neutrons.** Neutrons are the same weight as protons but do not have any electrical charge. Both the protons and neutrons adhere tightly together to form the dense, positively charged nucleus of the atom. Around this nucleus are two electrons so that, once again, the atom as a whole is neutral.

The structure of all the other atoms follows the same plan. Each of the other 101 kinds, from lithium (No. 3) to lawrencium (No. 103), consists of one more proton and one more electron than the atom just before it in the list (Fig. 1-2). Different atoms of a single element may have a different number of neutrons in the nucleus (these variant atoms are called **isotopes**), but this has little or no effect on the chemical properties of the atom.

The electrons surrounding the nucleus of each atom are restricted to certain definite "shells" or energy levels. These shells can be thought of as existing on the surface of a hollow sphere of a definite radius. Furthermore, there are limits to the number of electrons that can be present at any single energy level. At the innermost energy level only two can be present. Thus this level is filled in the helium atom. The next level can hold a maximum of eight electrons. These are

Fig. 1-2
Periodic table of the elements.

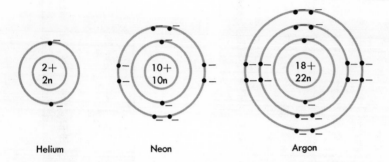

Helium   Neon   Argon

Fig. 1–3

The atomic structure of helium, neon, and argon. In each case there is a full complement of electrons in the outermost shell. Consequently, these elements do not react readily with other elements. They are called inert gases.

arranged in four pairs. The atom of the gas neon has eight electrons in this second shell (Fig. 1–3). The maximum number of electrons which can exist at any given level (n) can be computed from the expression

$$\text{maximum number} = 2n^2.$$

Thus, the third energy level can hold up to 18 electrons (nine pairs). How many electrons can exist at the fourth level?

There is one other important rule that describes the arrangement of electrons in the electrically neutral atom. This is that no more than eight electrons will be found in any energy level when it is the outermost one. For example, the third energy level can hold up to 18 electrons. It cannot hold more than eight, however, unless there are electrons in the fourth shell.

The arrangement of electrons in the atom plays a crucial role in its chemistry. Atoms are most stable when their outer shell contains its full quota of electrons. The so-called inert gases (e.g. neon and argon) have eight electrons in their outer shell.

Other atoms tend to gain or lose electrons until their outer shell also has the stable arrangement of the inert gases. Atoms with one, two, or three electrons in the outer shell usually lose these so that the next inner shell, with a more stable set of electrons, is left exposed. These atoms are called **electropositive.** The elements with atoms of this kind are the metals, such as sodium, magnesium, potassium, and calcium. Atoms with four, five, six, or seven electrons in their outer shell tend to add the necessary number of electrons to complete this shell. These atoms are **electronegative.** The elements with atoms of this kind are the nonmetals, such as carbon, nitrogen, oxygen, phosphorus, sulfur, and chlorine. The tendency for atoms to achieve a stable arrangement of electrons in the outer energy level explains why the various elements combine with one another the way they do. For example, table salt, sodium chloride, is made from equal numbers of sodium and chlorine atoms. The sodium atoms, with one electron in their outer shell, release this electron and thus gain stability. The chlorine atoms, with seven electrons in their outer shell, take on the electron from the sodium atoms and likewise gain stability (Fig. 1–4).

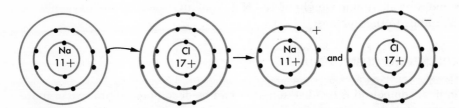

Fig. 1–4
Formation of sodium chloride. In the transfer of an electron from a sodium atom to a chlorine atom, each atom acquires an outer shell of 8 electrons and thus attains stability. Each also acquires an electrical charge. Charged atoms are called ions.

Our emphasis on the stable arrangements of electrons in an atom should not obscure the fact that unstable arrangements can exist. By adding energy to an atom, it is possible to "kick" electrons from a lower to a higher energy level. Such an atom is known as an "excited" atom. It is unstable. If left alone, the electron will soon return to its former energy level. In so doing, it will give back most of the energy that it had absorbed in the first place. When certain minerals are placed under ultraviolet light, they glow in brilliant colors. This is called **fluorescence.** It occurs because the electrons in the atoms of the mineral absorb energy from the ultraviolet light and are raised to a higher energy level. An instant later, however, they return to their former positions, giving up the absorbed energy in the form of visible light. (Fluorescent lamps work on exactly the same principle.) We will see later on that this ability of electrons to store energy is vitally important to the existence of life itself.

One other property of atoms that we should note is the fact that they have weight. Almost all of this is accounted for by the weight of the protons and neutrons in the nucleus. Hydrogen, with just a single proton in its nucleus, is the lightest atom. Uranium, with 92 protons and 146 neutrons in its nucleus, is the heaviest naturally occurring atom.

The weight of any single atom is, of course, infinitesimal when calculated in common units such as grams or ounces. To make calculations easier, we express the weights of atoms in terms of a special unit, the **atomic weight unit.** This unit is set arbitrarily as one-twelfth of the weight of a single atom of carbon (with six protons and six neutrons). Such a unit is very convenient because it makes the lightest element, hydrogen, have an atomic weight of 1. (Why?) It also makes individual protons and neutrons have a weight of 1. Thus the most common type of oxygen atom, which has eight protons and eight neutrons in its nucleus, has an atomic weight of 16. It is designated as O^{16}. It is only one of 3 so-called **isotopes** of oxygen found in the atmosphere. The others, however, make up less than 1% of the total number of oxygen atoms present. One of these isotopes has 8 protons and 9 neutrons in its nucleus, and thus its atomic weight is 17. Its symbol is O^{17}. The other, O^{18}, has 8 protons and 10 neutrons in its nucleus and thus an atomic weight of 18. Note that all the isotopes of oxygen have just 8 protons in the nucleus.

1-3 CHEMICAL COMPOUNDS

A chemical compound is a substance that can be decomposed into two or more simpler substances. If these simpler substances are also compounds, then they can be further decomposed. Ultimately, compounds can be decomposed into elements. The amounts of the elements in a given compound are not variable but are always present in definite proportions by weight. This is a reflection of the fact that the atoms of the elements in a compound are attached to one another in a precise way. Water is a compound. It is made up of the elements hydrogen and oxygen. The weight of the hydrogen in the compound is one-eighth the weight of the oxygen. This is understandable if two atoms of hydrogen (Atomic weight = 1) are attached to each oxygen atom (Atomic weight = 16) in the compound.

One of the most important things to remember about compounds is that their properties, such as color, taste, chemical activity, etc., are usually quite different from the properties of the elements of which they are made. The element sodium is poisonous and reacts explosively with water. The element chlorine was used as a poison gas during World War I. The compound sodium chloride, however, is absolutely vital to life activities. Its properties are quite different from those of either sodium or chlorine. Similarly, the properties of water are not the properties of the hydrogen or oxygen of which it is composed.

1-4 MOLECULES

Most of the compounds of special interest to biologists are made up of units called molecules. Each molecule consists of a precise arrangement of the atoms of the elements in the compound. It is this precise arrangement which accounts for the fixed ratio of elements in the compound. Because the constituent atoms have weight, the molecules have weight. The molecular weight (M.W.) is equal to the sum of the atomic weights of the atoms in the molecule. The molecular weight of water is 18. What is the molecular weight of carbon dioxide, CO_2?

It should be noted that many common gaseous *elements,* e.g. hydrogen, oxygen, nitrogen, and chlorine, are also composed of molecules. In these cases, only two atoms are present in the molecule, and the two atoms are alike. We designate these molecules as H_2, O_2, N_2, and Cl_2 respectively. What is the molecular weight of each of these molecules?

As we have seen, water is made up of molecules consisting of one oxygen atom to which are attached two hydrogen atoms. These atoms are held together by their tendency to achieve the most stable arrangement of electrons in their outer shells. Oxygen, with six electrons in its second shell, lacks two electrons to make a stable set of eight. Hydrogen, with one electron, needs a second to reach the stable arrangement of two electrons in its shell. By sharing electrons between them, each atom in a molecule of water gains a stable arrangement. Each hydrogen atom contributes an electron to the oxygen but, at the same time, electrons of the oxygen atom contribute to the formation of a pair for each of the hydrogen atoms. We can illustrate this by representing the nucleus (and all electron shells except the outer one) of the atoms by the symbol of the element and indicating the electrons in the outermost shell by dots or small crosses.

$$: \overset{..}{\underset{.\times}{O}} : H$$
$$H$$

Note that the oxygen atom now is surrounded by eight electrons and each hydrogen atom is surrounded by two. The process of sharing electrons causes the atoms to be linked or bonded together. This kind of bond is known as a **covalent bond.**

The number of electrons that a given atom has available to share in the making of covalent bonds sets a definite limit on its ability to combine with other atoms. Hydrogen, with just one electron to share, can only combine with one other atom. It is said to have a combining power, or **valence,** of one. Oxygen, needing to pair two of its electrons with two from elsewhere in order to achieve a complete set of eight, has a valence of 2. Nitrogen, with five electrons in its outer shell, needs three more electrons to form eight. Hence, it will share three of its electrons with three electrons from elsewhere. It has a valence of 3. The compound of nitrogen and hydrogen consists of one atom of nitrogen and three atoms of hydrogen.

$$H \cdot \overset{..}{N} : H$$
$$H$$

It is the substance ammonia. Carbon, needing four electrons to share with its own four electrons has a valence of 4. Its compound with hydrogen can be expressed as

$$H$$
$$H \overset{\times \cdot}{\underset{.\times}{C}} : H$$
$$\overset{.\times}{H}$$

This substance is a gas, methane, frequently produced in marshes.

Carbon atoms with a valence of 4 and oxygen atoms with a valence of 2 can also unite. In this case, two oxygen atoms unite with one carbon atom to form a molecule of the gas carbon dioxide, CO_2. The electron-dot representation of this molecule is interesting because it shows that not one, but two, pairs of electrons are shared between the respective atoms.

$$: \overset{..}{O} : \overset{\times}{\underset{\times}{C}} : \overset{..}{O} : \cdot$$

This arrangement is known as a **double** covalent bond. Note that it still provides for a complete set of eight electrons around each atom in the molecule.

1–5 FORMULAS

The electron-dot representations of the structures of molecules are a kind of formula. While they tell us a great deal about the arrangement of atoms in the molecule, they are a little complicated for regular use, particularly in considering large, complex molecules. Consequently, chemists have devised other, shorthand expressions for indicating the makeup of a compound.

One of these is the molecular formula. This formula simply lists all the kinds of atoms found in the molecule and, after each atom, includes a subscript giving the number of times (over one) that the atom appears in the molecule. The molecular formula for water is H_2O. The molecular formula for a molecule of the milk protein beta-lactoglobulin is approximately $C_{1864}H_{3012}O_{576}N_{468}S_{21}$. Note that the molecular formula simply gives the number of each of the various atoms that are found in the molecule. It does not tell us anything about the way in which these atoms are bonded together. In the case of beta-lactoglobulin, in fact, the arrangement of the atoms in the molecule is not yet known.

For many molecules we do know the actual physical arrangements of the atoms in the molecule. This is very important information, too, because the properties of a compound depend greatly on the actual way the atoms are linked together. For example, the common beverage alcohol, ethanol, has the formula C_2H_6O. However, there is another compound, an ether (related to the ether commonly used as an anesthetic), that also has the formula C_2H_6O. The difference in properties of these two compounds can be explained by the difference in the way the atoms are assembled in the molecule. We can show this difference by drawing a **structural formula** of each. In a structural formula we use a dash to represent each of the covalent bonds (which we showed above by pairs of dots and crosses). Using this system, we represent the ethanol molecule thus:

$$
\begin{array}{ccc}
 & H & H \\
 & | & | \\
H- & C- & C-OH \\
 & | & | \\
 & H & H
\end{array}
$$

On the other hand, the atoms of the ether molecule are attached to each other like this:

$$
\begin{array}{ccc}
H & & H \\
| & & | \\
H-C- & O- & C-H \\
| & & | \\
H & & H
\end{array}
$$

Note that each molecule has exactly the same number of atoms in it. Note also that in each case all of the valences are satisfied. Each of the two carbon atoms forms four bonds, the oxygen atom has two bonds, and each of the hydrogens forms only one bond. Two molecules, such as these, that have the same molecular formula but different structural formulas are called **isomers.** Isomers are found frequently in compounds containing carbon.

Even structural formulas are a little complex to use all the time. Consequently, chemists have devised condensed versions of them which still impart the essential information. Using condensed structural formulas, we can represent ethanol as CH_3CH_2OH or even C_2H_5OH. The ether is represented as CH_3OCH_3. Note that these condensed structural formulas have the great convenience of being able to be typed or printed in a single line. Note also that the distinguishing feature of each molecule, the "active group," is clearly shown. All alcohols have the —OH group. All ethers have the —O— arrangement.

You may well ask how chemists are able to determine the structural formula of a molecule. The first step is to determine its molecular formula. This is done by carefully decomposing a weighed amount of the compound into its constituent elements (or simple, inorganic molecules of these elements such as CO_2 and H_2O). From the weights of the breakdown products, it is possible to compute the relative numbers of the different atoms in the original molecule. A knowledge of the molecular weight gives the actual numbers and thus the molecular formula. (See question 10.)

There is no single next step in the procedure for determining the structural formula. The process depends upon the vast amount of knowledge about chemical properties of substances and "active groups" that has been accumulated by chemists for well over 100 years. Taking our earlier example, chemists know that the rules of valence will permit two carbon atoms, six hydrogen atoms, and one oxygen atom to be combined in just two ways. Through the years it has been found that alcohols react with other substances in certain ways, ethers in different ways. By finding out whether the chemical activity of our unknown compound is characteristic of the —OH group or the —O— group, we can hope to establish the true structural formula. The ultimate test comes, however, when we synthesize a compound, using another body of principles and rules, with the structural formula we have tentatively assigned to our "unknown." If the synthesized compound has the same physical and chemical properties as our "unknown," we can be quite confident that we do indeed understand the way the molecule is put together.

1–6 IONIC COMPOUNDS

Not all compounds are made of molecules. When atoms combine to form many compounds, the affinity of one of the atoms for electrons is so powerful that it is able to remove outer electrons completely from the other atom or atoms. Table salt is a good example of this. Chlorine, with its one missing electron, is so electron-attracting (**electronegative**) that it completely pulls away the one outer electron of the sodium atom (Fig. 1–4). Sodium, like all metals, does not have a great affinity for its outer electron (it is **electropositive**) so it gives its electron up readily. Having lost an electron, however, it now possesses a single positive charge (11 protons, but only 10 electrons). Similarly, the chlorine atom now has a single negative charge because of its additional electron (17 protons, 18 electrons). These charged atoms are called **ions.** The mutual attraction of opposite electrical charges holds the ions together by **ionic** bonds. The ions are not held together in pairs but are stacked in three-dimensional arrays (Fig. 1–5). Each sodium ion is held to six chloride ions (above, below, front, back, left, and right) while each chloride ion is, in turn, held by six sodium ions. The result of this stacking of ions is a crystal of table salt. At no place can one single out a pair of ions and refer to it as a molecule.

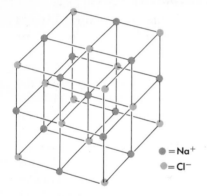

● $= Na^+$
● $= Cl^-$

Fig. 1–5

The lattice structure of a crystal of NaCl. The orderly stacking of Na^+ and Cl^- ions produces a crystal in the shape of a cube.

Since compounds that are ionically bonded contain no molecules, can we write formulas for them? Certainly, but in this case we are restricted to a formula that simply shows the ratios of the various positive and negative ions in the crystal. The formula NaCl shows that these two ions are present in a one-to-one ratio. Magnesium, with a valence of 2, forms crystals with two chloride ions for every one magnesium ion. The formula for this compound is $MgCl_2$. What would be the formula for aluminum chloride?

Because ionic compounds are not made of molecules, we cannot properly speak of the molecular weight of these compounds. We can easily arrive at the equivalent value, however, by adding up the atomic weights of the minimum ratio of ions in the crystal. We call this value the **formula weight.** The formula weight of sodium chloride is 58.5 (Na = 23, Cl = 35.5).

Ionic compounds are widespread in nature. Although we naturally associate them with the nonliving world of rocks and minerals, many ionic compounds are essential to life. Salts of sodium, potassium, calcium, chlorine, and other elements are found dissolved in the water of blood, cell fluid, etc. In this condition, the positive and negative ions become separated or **dissociated** from each other. The essential chemical properties of the substance are not changed in the process, however. Sodium ions (Na^+) and chloride ions (Cl^-) dissolved in water still retain the properties of table salt. They do not regain the poisonous properties of the neutral atoms from which these ions were made.

1-7 ACIDS

Even the molecules of covalently bonded compounds may dissociate into ions when they dissolve in water. Many covalent compounds containing hydrogen atoms are able to release one or more of them to nearby water molecules. For example, the gas hydrogen chloride is made up of molecules containing the two atoms held together by a shared pair of electrons, a covalent bond:

$$: \overset{..}{\underset{..}{Cl}} \text{ ⋮ } H$$

When this gas dissolves in water, the nucleus of the hydrogen atom (a proton) is attracted to one of the unshared electron pairs of a nearby water molecule. This leaves a negatively charged chloride ion

$$: \overset{..}{\underset{..}{Cl}} \text{ ⋮ }^{-}$$

and produces a positively charged **hydronium** ion, H_3O^+:

$$\begin{array}{c} H^+ \\ .. \\ H \text{ ⋮ } O \text{ : } \\ \cdot^\times \\ H \end{array}$$

The resulting solution is hydrochloric acid. It has a sour taste and other properties associated with acids. Any organic compound that possesses the **carboxyl** group

$$\begin{array}{c} O \\ \parallel \\ -C-OH \end{array}$$

will also dissociate into ions when placed in water. One of these will be a proton (H^+) which will immediately combine with a water molecule to form the hydronium ion (H_3O^+). Acetic acid has one carboxyl group and ionizes to form an acetate ion (CH_3COO^-) and a proton.

$$CH_3COOH + H_2O \rightarrow \dot{C}H_3COO^- + H_3O^+$$

Any compound that liberates protons in water solution is an acid.

1-8 CHEMICAL CHANGES

Chemical changes are changes in which new substances with new properties are formed. When hot sodium is surrounded by chlorine gas, the sodium bursts into flame. Table salt, sodium chloride, results. This is an example of a chemical change. The burning of coal is another example. Can you think of others?

All chemical changes involve a rearrangement of atoms (or ions). We can express these changes by means of an equation. To write a chemical equation we must first write the formulas of all the substances used in the chemical change. These are the **reactants.** Then we draw an arrow and write the formulas of all the substances produced by the chemical change, the **products.** When ethanol (ethyl alcohol) is exposed to air and the action of vinegar bacteria, it is changed chemically into vinegar (acetic acid) and water. We express this change by the equation

$$C_2H_5OH + O_2 \rightarrow CH_3COOH + H_2O$$

Note carefully that every atom present on the left side of the equation is accounted for on the right. If it is not, then the equation does not give a true picture of the chemical change. This is simply another way of saying that matter can neither be created nor destroyed. (We may neglect the special case of nuclear changes that occur in atom bombs, etc.)

In many chemical changes, the substances do not react in simple one-to-one ratios as they do in the production of acetic acid. When magnesium is burned in air, for example, two atoms of magnesium react with just a single molecule of oxygen. Two parts of the white powder magnesium oxide are produced. We express these relationships by placing the prefix 2 before the symbol for magnesium and the formula for magnesium oxide:

$$2Mg + O_2 \rightarrow 2MgO$$

You can easily carry out this chemical change yourself. Grasp a small piece of magnesium ribbon in a spring-type clothespin and hold the tip of the ribbon in a candle or bunsen burner flame. The heat of the flame will first soften the ribbon. Then the ribbon will begin to burn. In so doing, it gives off a good deal of heat and a brilliant light. (Do not stare at the light.) When the burning is completed, a white powder, magnesium oxide, remains.

An equation can tell us even more than what rearrangement of atoms has taken place. It can also tell us what the relative weights of the reactants and products of the chemical change are. From the equation given above, you might think that 1 gram (gm) of ethanol and 1 gm of oxygen would produce 1 gm of acetic acid and 1 gm of water. Such is not the case, however. The equation tells us that one molecule of ethanol reacts with one molecule of oxygen. It implies that likewise a billion molecules of one will react with exactly a billion molecules of the other. Thus, the 1:1 ratio expresses the number of each type of molecule in the reaction. You can quickly see by computing the molecular weights of these two molecules that the ethanol molecule is somewhat heavier (M.W. = 46) than the oxygen molecule (M.W. = 32). If we want a given quantity of ethanol to react completely with oxygen, we must supply one molecule of oxygen for every molecule of ethanol. Counting molecules is not practical. Weighing substances, on the other hand, is a very efficient technique. To make sure that equal numbers of ethanol and oxygen molecules react, we must bring about a reaction between weights of these two substances which are in the ratio of 46:32.

Another example may help make this point clear. As a biology student, you might be interested in comparing the effects of sodium chloride and potassium chloride solutions on the rate of beating of an isolated frog heart. At first you might plan to prepare one-percent solutions (1 gm of the salt in 99 gm of water) of each substance. This would be a poor technique, however. The formula weights of these two salts are 58.5 and 74.5 respectively. A one-percent solution of sodium chloride would thus contain almost half again as many ions as a one-percent solution of the potassium chloride. (Why?) In such a case, you could not properly relate the response of the frog heart to the kind of metal ion applied.

A much better technique is to make up the salt solutions so that equal volumes of each of the two solutions contain equal numbers of the ions. To do this, we simply weigh out 58.5 parts of NaCl to every 74.5 parts of KCl. Then these two quantities must each be added to enough water to make the two resulting solutions of exactly equal volume. Once this is done, you know that, drop for drop, the two solutions have exactly the same concentration in terms of numbers of ions present.

In trying to weigh out two substances in the ratio of 58.5 to 74.5, what could be more simple than to weigh out 58.5 gm of one and 74.5 gm of the other? In so doing, you would have weighed out a quantity of each substance called a **mole.** If you then add enough water to each substance to make exactly one liter of solution, you will have made a 1-molar (1-M) solution of each salt. [A specially graduated flask is a useful device in which to do this. After the substance is added to the flask, enough water is added to bring the volume up to the engraved line on the neck of the flask. The volume of solution is then exactly one liter (Fig. 1–6).] A 1-M solution of these salts would undoubtedly be too strong for the experiment on the frog heart. It would probably be better to make up a liter of each solution containing 5.85 and 7.45 gm respectively. Such solutions would be designated one-tenth molar (0.1M) solutions. Drop for drop, these two solutions would also contain exactly the same number of ions because they are of the same molarity.

Looking back at our earlier example, we are now in a position to estimate how much acetic acid and how much water would be produced as a result of reacting one mole of ethanol (46 gm) with one mole of oxygen (32 gm). One mole of each substance would be produced, namely 60 gm of CH_3COOH and 18 gm

Fig. 1–6

A volumetric flask. When filled to the etched line, it contains exactly one liter. Such a flask is used to prepare solutions of precise molarity.

of water. Note, also, that the total weight of the products (60 gm + 18 gm = 78 gm) is exactly equal to the total weight of the reactants (46 gm + 32 gm = 78 gm). This, once again, illustrates the principle that during chemical changes, matter is neither created nor destroyed. (In practice, when organic compounds react, one seldom gets a mole of each product for each mole of reactant, i.e. 100% efficiency. This is not because any matter is lost. It is simply because other reactions, "side" reactions, go on simultaneously and produce unwanted products.)

1–9 TYPES OF CHEMICAL CHANGE

There are several types of chemical change. Two of these are of special importance to the beginning biology student. These are ion-exchange reactions and oxidation-reduction (REDOX) reactions.

1.   Ion-Exchange Reactions

When hydrogen chloride gas (HCl) is dissolved in water, a solution of hydrochloric acid is produced.

$$HCl + H_2O \rightarrow H_3O^+ + Cl^-$$

As explained earlier, this reaction simply involves the transfer of an ion, a proton (H^+), from one molecule to another. There has been no change in the arrangement of the electrons. The HCl molecule donated the proton and, like all proton donors, is called an acid. The water molecule, which accepts the proton, acts as a **base.** It is not a very strong base, though, because it will readily give its proton up to other proton acceptors. One of the most powerful of these is the hydroxyl ion. This ion is produced when substances called hydroxides (e.g. NaOH) are dissolved in water. The OH^- ion has a greater affinity for protons than does the water molecule. Consequently, the proton shifts from the hydronium ion, now the acid, to the hydroxyl ion, the base.

$$H_3O^+ + OH^- \rightarrow 2H_2O$$

If one adds equal volumes of an acid solution and a basic solution (of the same concentration of reacting ions) together, they will react completely in this way. This process is called **neutralization** because the resulting solution has neither the properties of an acid nor the properties of a base. In fact, the solution produced by this particular neutralization is simply salt water.

$$H_3O^+ + Cl^- + Na^+ + OH^- \rightarrow 2H_2O + Na^+ + Cl^-$$

The OH^- ion is a very powerful proton acceptor and thus a very strong base. It is never found in high concentrations in living organisms. Probably the most important base in living tissue is the bicarbonate ion, HCO_3^-. This base readily neutralizes excess acid that may be formed by cell metabolism. Perhaps you know of people who take sodium bicarbonate to neutralize the hydrochloric acid in their stomachs in an attempt to alleviate "acid indigestion."

$$H_3O^+ + Cl^- + Na^+ + HCO_3^- \rightarrow H_2O + H_2CO_3 + Na^+ + Cl^-$$

In summary, when any molecule or ion gives up a proton, it is acting as an acid. When any molecule or ion takes on a proton, it is acting as a base. Any acid and any base will neutralize each other if each solution contains the same number of ions.

2. Oxidation-Reduction Reactions

Oxidation-reduction reactions differ from ion-exchange reactions in that a transfer of *electrons* is always involved. You remember that when sodium burns in chlorine (Fig. 1–4) each sodium atom loses an electron to a chlorine atom. The atom that loses the electron (sodium) is said to have been **oxidized.** The atom that gains the electron (chlorine) is said to have been **reduced.** The two processes are completely linked. Whenever any substance gives up electrons, it is oxidized. The substance to which it gives its electrons is called the oxidizing agent. Note, however, that any substance, in the process of acting as an oxidizing agent, is itself reduced. Because oxidations always go hand in hand with reductions, we refer to these reactions as REDOX reactions.

You may wonder why the reaction between sodium and chlorine is called an oxidation when oxygen plays no part in it. The answer is simply a matter of history. Oxygen has a great affinity for electrons and is thus one of the best and most common oxidizing agents known. Because of its effectiveness in this role, it has supplied the name for all reactions of this type.

The burning of magnesium is an example of an oxidation that actually does involve oxygen. The outer two electrons on the magnesium atom (an electropositive element—see Section 1–6) are taken on by an atom of oxygen (a strongly electronegative element), thus completing its outer shell of eight electrons. All strongly electronegative elements like oxygen and chlorine are good oxidizing agents. They are very effective at removing electrons from other substances. Conversely, electropositive elements like sodium and magnesium are easily oxidized. They give up their outer electrons easily. They are good reducing agents.

Some biological oxidations involve the addition of oxygen to the molecule being oxidized. In the reaction between ethanol and O_2, one atom of oxygen joins the molecule. Although the added oxygen atom does not remove electrons sufficiently far to form ions (as in the case of NaCl and MgO), it nevertheless does attract two electrons away from the carbon atom and much closer to itself (Fig. 1-7).

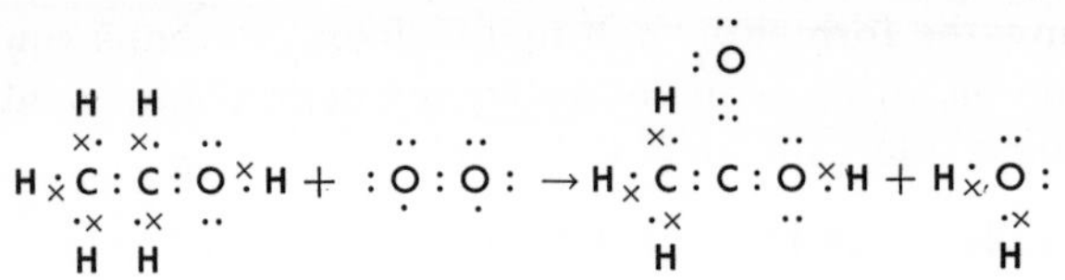

$$
\begin{array}{ccc}
& & :\ddot{O} \\
H\ H & & H\ \vdots\vdots \\
\overset{\times\cdot}{\ }\overset{\times\cdot}{\ }\ \cdots & \cdots\ \cdots & \overset{\times\cdot}{\ } \\
H\underset{\times}{C}:C:O\overset{\times}{\vphantom{.}}H + & :\underset{\cdot}{O}:\underset{\cdot}{O}: \rightarrow & H\underset{\times}{C}:C:O\overset{\times}{\vphantom{.}}H + H\underset{\times}{\vphantom{.}}\ddot{O}: \\
\overset{\cdot\times}{\ }\overset{\cdot\times}{\ }\ \cdots & & \overset{\cdot\times}{\ }\ \cdots & \overset{\cdot\times}{\ } \\
H\ H & & H & H
\end{array}
$$

Fig. 1-7

The oxidation of ethanol. Oxidation has occurred both by the addition of an oxygen atom and by the removal of hydrogen atoms. In each case, electrons are moved away from carbon atoms.

The most common kind of biological oxidation is accomplished by the removal of hydrogen atoms from a substance. The oxidation of ethanol also exhibits this kind of oxidation. Two hydrogen atoms, each with its single electron, are removed by the second oxygen atom. These unite to form a molecule of water.

REDOX reactions in which the electrons shift from a less electronegative atom to a more electronegative one proceed spontaneously and liberate energy in the form of heat. The burning of the magnesium strip is an excellent illustration of this. The reason that energy is liberated is that the electrons are moving from a less stable but energy-rich arrangement to a more stable but energy-poor one. The products of the reaction contain less potential ("stored") energy than the reactants. In the process of the reaction, the difference is liberated as heat and light.

We heat our homes and run electric generators and automobile engines with the energy released by the oxidation of fuels (e.g. coal, oil, and gasoline) which all contain substantial quantities of potential energy. The products of their oxidation, CO_2 and H_2O, contain far less potential energy. We run our own life activities on the energy released by oxidizing the foods we eat. This process of biological oxidation is called **respiration.**

REDOX reactions in which the shift of electrons is from a more electronegative to a less electronegative atom require an input of energy. The energy is required to force electrons away from the more electronegative (greater electron affinity) atom to the less electronegative (less electron affinity) atom. The most important reductions in the biological world are those carried out by green plants during the process of photosynthesis. Light supplies the energy for these. Some of the light energy becomes stored as the potential energy of the foods produced. In fact, it is this stored energy that is released again when foods and other fuels are oxidized. The oxidation of a given quantity of food, or any other energy-rich substance, releases just the amount of energy that was consumed during the formation of that substance.

Reductions in living organisms are often accomplished by the addition of hydrogen atoms (each with its electron) to a substance. They may also be accomplished simply by the transfer of electrons alone. Some of the metallic elements found in living things have variable valences and play an important part in this. Iron atoms may be quite easily oxidized by the removal of the two electrons in the outer shell.

$$Fe - 2e^- \rightarrow Fe^{++}$$

A third electron, this one from the shell beneath, can also be removed by oxidation.

$$Fe^{++} - 1e^- \rightarrow Fe^{+++}$$

The Fe^{+++} ion is a fairly good oxidizing agent itself, however. It can remove an electron from some other substance (which is thus oxidized) and become the Fe^{++} ion once again. Then it can turn right around and act as a mild reducing agent, donating its recently acquired electron to still another substance. This ability of iron to pick up and pass on electrons depends upon its variable valence. Iron and certain other metallic elements play a major role in the REDOX reactions carried out in living things.

1–10 RATE OF CHEMICAL REACTION

Several factors affect the speed at which chemical changes take place. One of the most important of these is simply the **nature of the reacting materials.** You can well imagine that a vigorous, energy-releasing oxidation, with electrons passing from electropositive to electronegative atoms (as in the burning magnesium ribbon), proceeds much more rapidly than the reverse.

Temperature also affects the rate of chemical reaction. A great deal of the energy liberated by the burning magnesium ribbon is in the form of heat. This heat energizes still-unoxidized magnesium atoms so that they combine even more quickly with oxygen atoms. As a general rule, the speed at which chemical reactions take place doubles with every 10° C rise in temperature. This is true not only for reactions in the test tube but also, within limits, for reactions in living things.

The **concentration** of reacting substances is a third factor in regulating the speed at which chemical changes occur. If the reacting molecules are far apart, their opportunities for colliding and exchanging atoms and electrons are limited. On the other hand, if they are packed closely together, there is a much greater probability of their colliding and of a resulting chemical change.

The reaction $2Mg + O_2 \rightarrow 2MgO$ involves materials that are brought together whenever a piece of magnesium ribbon is exposed to the air. Why doesn't the reaction proceed spontaneously? Why is it necessary to *add* energy (heat) to the magnesium ribbon when the reaction between magnesium and oxygen yields such abundant amounts of energy? The answer lies in the need to activate the magnesium atoms. Magnesium atoms will not transfer their electrons to oxygen atoms unless they are activated by the input of energy. At ordinary temperatures, very few atoms become sufficiently active to combine with oxygen. Those that do, however, liberate far more energy in the course of reacting than it took to activate them (Fig. 1–8). Nevertheless, because of the small numbers involved, the resulting heat is liberated so slowly that it does not warm up the metal. Consequently the oxidation of magnesium proceeds very slowly. (The slow oxidation of metals is called corrosion or, in the case of iron, rusting.)

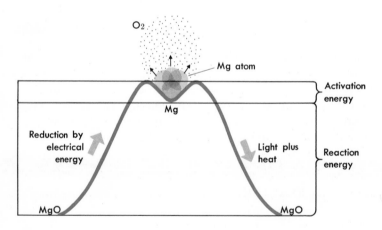

Fig. 1–8
A model showing the energy changes in a REDOX reaction.

At higher temperatures, a sufficiently large number of magnesium atoms become activated and react with oxygen so that the resulting heat raises the temperature of the metal still further. This activates still other atoms and the reaction becomes self-sustaining. Heat is the most common way in which we supply the energy of activation for chemical reactions. The sparkplug on a gasoline motor and the blasting cap in a stick of dynamite are two examples. Can you think of others?

1–11 CATALYSIS

The use of heat to supply activation energy is not too suitable for use by living things. High temperatures are quite destructive to life, as we all know. Fortunately, there is a way around the problem. This is to reduce the energy of activation needed by the use of a suitable **catalyst.** A catalyst is a chemical substance that speeds up the rate at which a given chemical change takes place without becoming permanently altered in the process. Many catalysts accomplish this by temporarily uniting with one of the reacting substances and lowering the amount of activation energy needed. When the reaction is completed, the catalyst is released unchanged and ready for re-use. You can demonstrate catalysis easily with a small quantity of a 3% solution of hydrogen peroxide (H_2O_2)— available in any drug store—and a pinch of the black compound manganese dioxide (MnO_2). At room temperature and atmospheric pressure, and especially when exposed to light, hydrogen peroxide breaks down to form water and oxygen ($2H_2O_2 \rightarrow 2H_2O + O_2$). (This is why the manufacturer puts the solution up in a brown glass bottle and recommends that it be kept in a cool place with the cap screwed on tightly.) When the solution is poured into a glass, a few bubbles of O_2 soon form. Adding a pinch of MnO_2 greatly increases the speed of the reaction, however. A froth of O_2 bubbles forms immediately (Fig. 1–9). Thus catalysis takes its place as one of the factors that speed up the rate of chemical reactions.

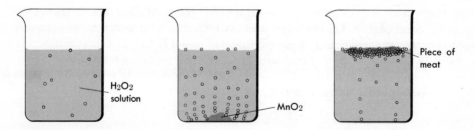

Fig. 1–9

Catalysis. Left: A solution of H_2O_2 spontaneously decomposes, releasing a few bubbles of oxygen. Center: The action is greatly speeded by the addition of the inorganic catalyst manganese dioxide. Right: Cells, in a piece of meat in this case, contain an enzyme, catalase, that likewise hastens the action.

Living cells have catalysts, too, and thus are able to carry out their chemical activities quickly but at safe temperatures. These catalysts are called **enzymes.** It is thought that every chemical reaction occurring in living things (there may be thousands of them) requires the presence of a specific enzyme. Most cells even have an enzyme, called *catalase*, that decomposes H_2O_2 just as MnO_2 does. It is responsible for the frothing which occurs when hydrogen peroxide is poured into an open wound. You can, however, demonstrate the action of the enzyme under more pleasant circumstances by adding a piece of fresh meat to a container of H_2O_2. The meat will quickly float to the top, buoyed up by the bubbles of O_2 that form on its surface.

1-12 REVERSIBLE REACTIONS

In Section 1-9, you learned that the hydronium ion and the bicarbonate ion react to form water and carbonic acid according to the equation:

$$H_3O^+ + HCO_3^- \rightarrow H_2O + H_2CO_3$$

The carbonic acid (H_2CO_3) that is formed is unstable and decomposes into a molecule of carbon dioxide and a molecule of water:

$$H_2CO_3 \rightarrow H_2O + CO_2$$

The complete reaction is thus

$$H_3O^+ + HCO_3^- \rightarrow H_2O + H_2CO_3$$
$$\downarrow$$
$$H_2O + CO_2$$

In a test tube, the carbon dioxide is released into the air, and all the hydronium ions are eventually neutralized. In a closed container, the situation is somewhat different. The carbon dioxide that is produced remains in contact with the water. As the reaction continues, the concentration of CO_2 increases steadily. Soon carbon dioxide molecules are colliding with water molecules and *back-reacting* to form H_2CO_3. Some of this additional H_2CO_3 dissociates in water to form H_3O^+ and HCO_3^-. Reactions in which the products can back-react to form the reactants again are called **reversible.** We use double arrows to indicate the reversible nature of the reaction.

$$H_3O^+ + HCO_3^- \rightleftharpoons H_2O + H_2CO_3$$
$$\upharpoonleft\downharpoonright$$
$$H_2O + CO_2$$

In a closed system, the forward and reverse reactions eventually reach **equilibrium.** At this time, the rate of the forward reactions exactly equals the rate of the reverse reactions. This does not necessarily mean that the concentrations of products and reactants are equal. At equilibrium in the second reaction of our example, in fact, the concentration of CO_2 and H_2O may be some 1000 times greater than that of H_2CO_3. For every molecule of CO_2 that reacts with H_2O, however, a molecule of H_2CO_3 decomposes into H_2O and CO_2. Similarly, at equilibrium, for every hydronium ion that reacts with a bicarbonate ion, a molecule of H_2CO_3 dissociates into its ions.

Equilibria can be disturbed in several ways. Changing **temperature** usually favors one reaction over the other, and thus brings about a shift in the equilibrium. **Adding** fresh **reactants** to the system favors the reaction that consumes the reactants. In our example (see the previous equation) addition of fresh acid displaces the equilibrium to the right, favoring the formation of increased amounts of carbon dioxide. **Removing a product** accomplishes the same thing. As we noted above, in an open test tube the carbon dioxide leaves the reaction mixture and no back reaction is possible. Under these conditions, the forward reaction (neutralization) is said to go to completion.

What about catalysts? Although catalysts have dramatic effects on the rates at which reactions occur, they do not displace reversible reactions in either direction. This is because both the forward and back reactions are speeded up to the same extent. Catalysts thus enable equilibrium to be reached more quickly but do not favor one reaction over its reverse.

The majority of the chemical reactions that occur in living organisms are reversible. Which direction is favored depends upon changes in the factors mentioned above. Proper functioning in the organism depends, in turn, upon the reaction's proceeding in a particular direction at a particular time.

1–13 MIXTURES

A mixture is a material composed of two or more pure substances each of which retains its own characteristic properties. When sand is stirred into water, a mixture results. The properties of the sand and the water are unchanged in the process. (You remember that compounds have properties quite different from the properties of their constituents.) Another way in which mixtures differ from compounds is that their composition can be varied. You can mix varying amounts of sand and water together. Magnesium oxide, on the other hand, always contains magnesium atoms and oxygen atoms in the ratio of 24:16 parts by weight.

A mixture like that of sand and water is called a **suspension.** After a period of time, the sand particles (crystals of SiO_2) settle to the bottom of the container under the influence of gravity. Quite a different sort of mixture is a **solution.** In a solution, individual molecules or ions of a substance (the **solute**) become suspended in a liquid medium (the **solvent**). These molecules or ions are so small (less than 1 mμ*) that they never settle out under the influence of gravity. Their own motion counteracts the gravitational force acting on them. When a salt (NaCl, KCl, etc.) or sugar is added to water, a solution results. The mixture is homogeneous, that is, even under the microscope it is perfectly uniform in appearance. While it is easy to separate sand and water by filtering through a cone of paper (Fig. 1–10), solutions cannot be separated in this way. The sizes of the solute molecules and ions in a solution are approximately the same as the size of the solvent molecules and pass through the pores of the filter paper just as

---

* A table explaining the metric system of measurement appears on page 194.

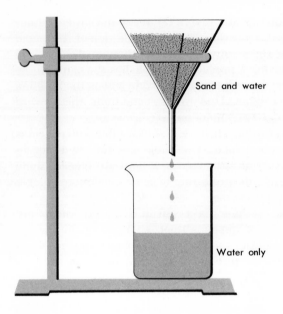

Fig. 1–10

Filtration. Any particles, such as sand, too large to pass through the pores in the paper can be separated in this way from smaller molecules such as those of water.

Sand and water

Water only

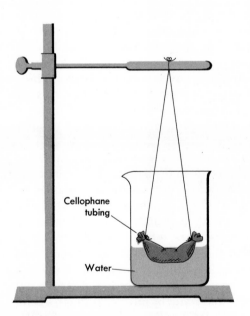

Cellophane tubing

Water

Fig. 1–11

Dialysis. The pores in a piece of cellophane are so small that macromolecules (e.g. starch, egg white) are retained inside while small molecules and ions (e.g. glucose, Na^+, Cl^-) pass out readily into the surrounding water. Dialysis is the separation of macromolecules from small molecules and ions.

easily. Although other liquids can act as solvents, water is the major solvent in living things.

The student of biology should also be familiar with a third kind of mixture, the **colloid.** Colloids are suspensions of particles that are larger than those in true solutions but still too small to settle out under the influence of gravity. These particles may consist of single, large molecules (1–100 mμ) or solid clumps of smaller molecules. They are large enough to be seen under very high magnifications. They cannot be filtered out by ordinary filter paper because the pores in the paper are too large to stop their passage. Cellophane, however, has pores sufficiently small to separate colloidal particles from smaller molecules (Fig. 1–11). Egg white is a familiar colloid.

Fig. 1–12

Separating macromolecules by column chromatography. A mixture of protein molecules in solution drips onto the top of the chromatographic column on the right. The packing material in the column separates the molecules according to their size. As the isolated proteins leave the bottom of the column, they are collected in separate test tubes. (Courtesy Instrumentation Specialities Company, Inc.)

The biologist is especially interested in colloids because a large proportion of the molecules found within living cells are **macromolecules,** that is, molecules so large that they form colloids. As we will see in the next chapter, there are several categories of macromolecules within the cell and for some of these, e.g. proteins, a wide variety of specific types. The successful analysis of the chemistry of cells has depended in large measure on the development of techniques for separating these macromolecules from each other. Slight differences in size, shape, electrical charge, and solubility in different solvents must be exploited to accomplish these separations. Figure 1–12 shows a device, called a chromatographic column, used to separate closely related macromolecules (in this case, proteins) from each other.

We now turn our attention to the general topic of the molecules from which living things are made.

EXERCISES AND PROBLEMS

1   Which of the following substances are composed of molecules: (a) oxygen, (b) water, (c) sodium chloride, (d) glucose, (e) steel?

2   Distinguish between an atom and an ion.

3   What ion is produced by all acids?

4   A mole of methane (CH_4) weighs how many grams?

5   Give an example of a strongly electropositive element.

6   Write electron-dot formulas for the molecules of (a) water, (b) ammonia, (c) methane, (d) ethane (CH_3CH_3).

7   Summarize the differences between mixtures and compounds.

8   Distinguish between organic and inorganic compounds.

9   Which of the following elements would you expect to react with chlorine: (a) hydrogen, (b) neon, (c) sodium, (d) fluorine, (e) calcium, (f) carbon?

10   Chemical analysis shows that the ratio of C atoms to H atoms in a substance is 1 to 3. The molecular weight of the substance is 30. What is the molecular formula?

11   Show by means of electron-dot formulas the combining of a lithium atom (3 electrons) with a fluorine atom (9 electrons).

12   A student needs 200 ml of 0.1 M NaOH solution. What weight of solid is used?

13   What is the molecular weight (MW) of acetic acid (CH_3COOH)?

14   When methane (CH_4) burns in oxygen, carbon dioxide and water are the products. Write a balanced equation to represent this chemical change.

15   What volume of 0.4 M hydrochloric acid will neutralize 200 ml of 0.2 M NaOH solution?

16 What is the most abundant metal in the crust of the earth?

17 Write a structural formula for butane (C_4H_{10}).

18 Write a different structural formula for butane.

19 What term is used to describe these two compounds?

20 Distinguish between electron, proton, and neutron.

21 Write an electron-dot formula for carbon tetrachloride (CCl_4).

22 Distinguish between isomers and isotopes.

REFERENCES

Additional information on the topics discussed in this chapter can be found in any good textbook on introductory chemistry.

A small book that covers topics in chemistry and physics, selected and developed with the needs of biology students in mind, is:

Baker, J. J. W., and G. E. Allen, *Matter, Energy, and Life: An Introduction for Biology Students,* Addison-Wesley, Reading, Massachusetts, 1965.

CHAPTER 2 THE CHEMICAL BASIS OF LIFE: THE MOLECULES OF LIFE

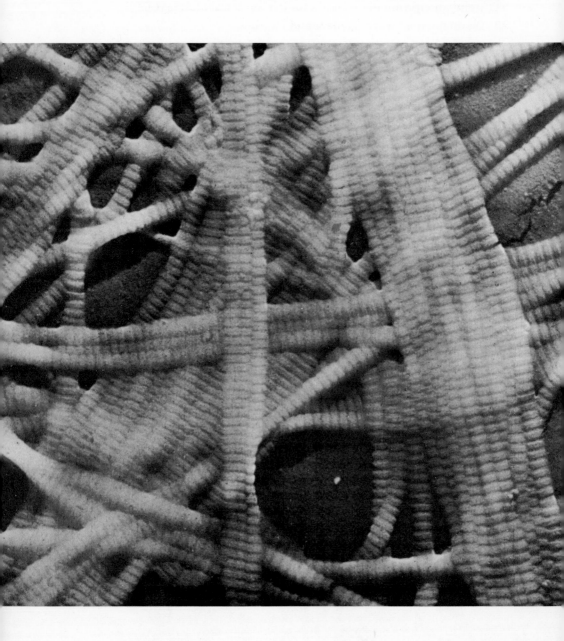

The same basic chemical principles apply to matter found in living as well as nonliving things. Matter found in living things does, however, have certain distinguishing characteristics. One is that the number of different elements found in living matter is considerably smaller than the number found in non-living matter. Furthermore, living things are largely composed of carbon, hydrogen, nitrogen, and oxygen. Of these elements, only oxygen is plentiful in the nonliving world.

Perhaps the one element which, more than any other, is characteristic of living matter is carbon. Because of the ability of carbon atoms (1) to unite with one another and thus form long chains and (2) to form double covalent bonds, they participate in the formation of an almost infinite variety of molecules. Such variety is essential for such an incredibly complex organization of matter as a living thing. It is no wonder, then, that molecules containing carbon form the very basis of life itself. In fact, the chemistry of carbon compounds is called **organic chemistry** because these compounds are almost exclusively associated with life processes.

Molecular composition of the human infant	
Water	66%
Proteins + nucleic acids	16%
Fats and lipids	12.5%
Ash (minerals)	5%
Carbohydrates	0.5%
Vitamins	trace

Fig. 2–1
In the human infant, water represents almost two-thirds of its weight.

It is fortunate for our understanding of the organization and functioning of living things that the almost unlimited numbers of organic compounds do fall into a reasonable number of distinct categories. As far as quantity is concerned, organisms are composed mainly of water, proteins, fats and other lipids, carbohydrates, nucleic acids, and minerals (Fig. 2–1). Of these, water and minerals are inorganic. The other substances are all organic.

2–1 CARBOHYDRATES

Carbohydrates are essential to life. Almost all organisms use them as a fuel, exploiting their rich supply of potential energy to maintain life. Carbohydrates also serve as an important structural material in some animals and in all plants.

◄ Filaments of collagen, a protein, 31,000 X. (Courtesy Dr. Jerome Gross.)

Carbohydrates are made of molecules containing carbon, hydrogen, and oxygen atoms with the ratio of hydrogen atoms to oxygen atoms always 2:1. In view of this, can you imagine how the word carbohydrate came to be used for these compounds?

Many carbohydrate molecules are extremely large, possessing molecular weights of 500,000 or more. Fortunately, these **macromolecules** are composed of simpler, repeating units, and this makes them relatively easy to study. The repeating units are called sugars.

1. Sugars

The most important sugar is **glucose.** It has the molecular formula $C_6H_{12}O_6$, but as is so often the case in organic chemistry, the molecular formula is not sufficient to describe the molecule fully. There are other common sugars that have precisely the same molecular formula and thus are **isomers** of glucose. Glucose can, however, be distinguished from these by its structural formula (Fig. 2–2). Five of the six carbon atoms and one of the oxygen atoms form a flat ring. The hydrogen atoms and other groups extend above and below the plane of this ring.

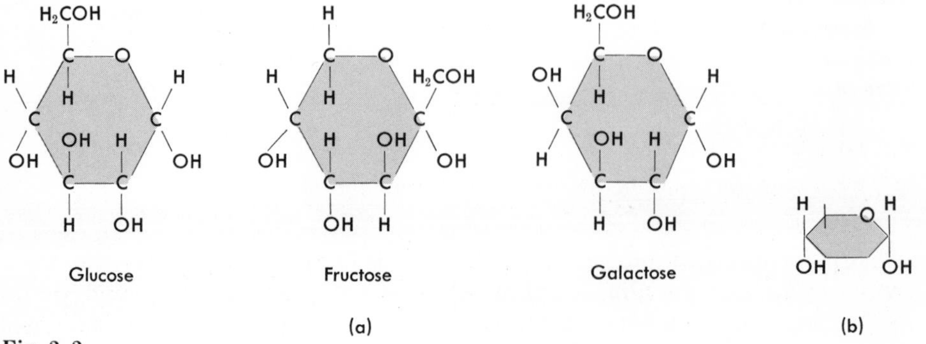

Glucose   Fructose   Galactose

(a)   (b)

Fig. 2–2

(a) Structural formulas of glucose and two of its isomers. (b) Simplified version of the glucose molecule.

The great importance of glucose to life is that it serves as the basic, transportable form of fuel for the organism. It is soluble and thus easily transported by body fluids. In humans, glucose is actually referred to as "blood sugar." Its concentration in the blood must be maintained within narrow limits or serious disturbances will result. Glucose is the starting material for cellular respiration, the process from which most organisms derive the energy they need to run their life activities.

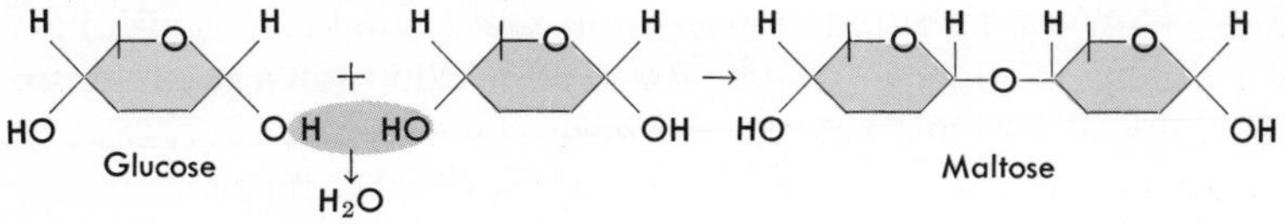

Fig. 2–3

Synthesis of maltose.

Two glucose units can be linked together by an oxygen bridge between them (Fig. 2–3). This is formed by the removal of a molecule of H_2O. A molecule of **maltose** results. It is a double sugar or **disaccharide,** with the molecular formula

$$C_{12}H_{22}O_{11}   (2C_6H_{12}O_6—H_2O).$$

Maltose is produced when starch is digested (see below).

The sugar that you put in your coffee or on your breakfast cereal, **sucrose,** is also a disaccharide. It is formed by the linking together of a glucose unit and a unit of its isomer **fructose.** (What, then, is the molecular formula of sucrose?) Our chief source of sucrose is sugar cane and sugar beets. Most higher plants transport carbohydrate as sucrose. It is very soluble in water and can thus be easily transported in sap.

2.   Starches

Starches are large carbohydrate molecules made up of chains of glucose units (Fig. 2–4). We use the term **polymer** to describe chainlike molecules of this sort. The "links" in the chain are **monomers.** Most natural starches contain both

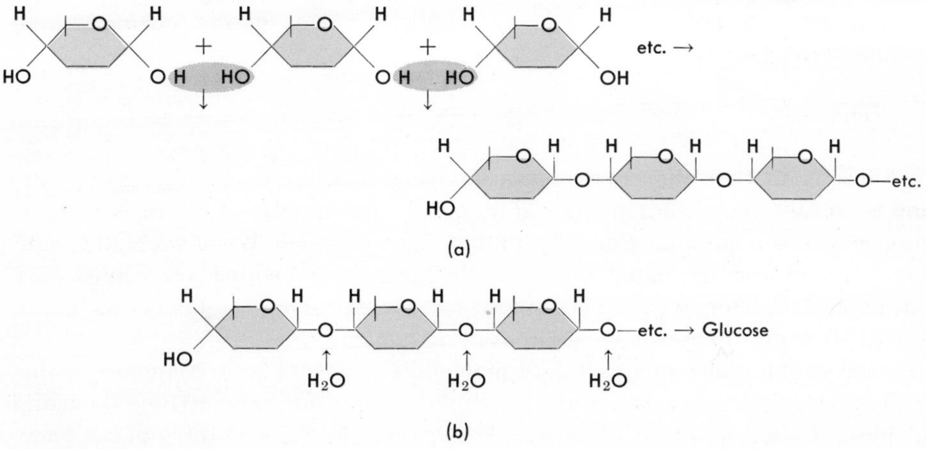

(a)

(b)

Fig. 2–4

(a) Synthesis of starch. (b) Hydrolysis of starch.

unbranched chains of several hundred glucose units and branched chains total-
ling more than a thousand glucose units. Because of the many sugar units
present in starches, they are called **polysaccharides.**

Starches are important because they serve as a storage form of sugar.
Surplus sugar can be converted into starch, which is insoluble, and stored. This
takes place in both plants and animals. Man's greatest source of carbohydrates
is plant starch. Most of the world's population satisfies its energy needs with
the starches of rice, wheat, corn and potatoes. Before starches can be absorbed
into the body, however, they must be digested. This simply means that the long
chains are broken back down into their sugar links. This process requires both
water and starch-digesting enzymes called **amylases.** With the aid of amylases,
water molecules enter at each oxygen bridge, re-forming free sugar molecules
(Fig. 2–4). The process of breaking up a molecule by inserting water is called
hydrolysis.

Amylases hydrolyze starches into the disaccharide maltose. Another enzyme, maltase, hydrolyzes the maltose molecule into two glucose molecules.

When man takes in more glucose than he needs, he converts some of the
excess to starch. This starch, as well as that produced by other animals, differs
somewhat in its properties from plant starch. For this reason it is often given a
special name, **glycogen.** The main difference between glycogen and plant starch
lies in the pattern of branching. There seem to be no straight chain molecules in
glycogen like those in plant starches.

Man stores glycogen in his muscles and in his liver. In these locations, it is
quickly available for breakdown into glucose and subsequent energy production.
Glycogen is not, however, the major storage form of energy in man. We depend
on fat deposits for most of our energy reserves. There are some animals, though,
that do rely almost entirely on glycogen stores. These are apt to be animals such
as mussels (mollusks) that regularly undergo periods of oxygen shortage. The
advantages of glycogen as a source of energy when oxygen is lacking will be
discussed in Chapter 5.

3. Cellulose

Cellulose is another important polysaccharide. It is universally found in plants
and is, in fact, their chief structural material. The rigidity of plants is a conse-
quence of the large quantities of cellulose they produce. Wood is chiefly cellulose. Cotton and the paper upon which this book is printed are almost pure
cellulose. The extensive use of cellulose as a structural material in plants makes
it without doubt the most prevalent organic molecule on earth.

Like starch molecules, cellulose molecules consist of long, branched chains
of glucose units. A single cellulose molecule may contain over three thousand
of these. Cellulose differs from starch, however, in the way the glucose molecules are attached to one another. In starches, the glucose molecules are all
oriented in the same fashion. In cellulose the orientation of the glucose units
alternates from one glucose molecule to the next (Fig. 2–5).

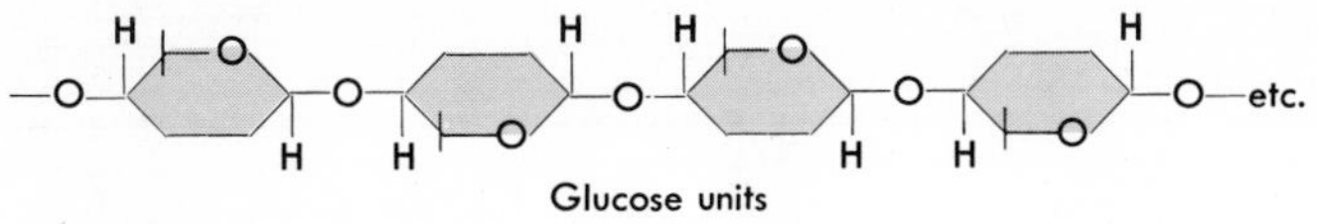

Glucose units

Fig. 2–5

Structure of cellulose. Compare its structure with that of starch (Fig. 2–4).

This difference in molecular structure is sufficiently great that amylases are unable to hydrolyze cellulose. Cellulose can be digested by **cellulases,** which are produced by certain bacteria, protozoans, and a few invertebrates such as terrestrial snails and some insects. What of the cows, rabbits, and termites which seem to thrive on cellulose? In each of these cases, the cellulose digestion is accomplished by microorganisms living within the animal's digestive system. The rumen of cows and the caecum of rabbits are large pouches in which food is stored while bacteria digest cellulose with their cellulases. Both the bacteria and their mammalian host then absorb the sugars produced. Termites also depend upon microorganisms to carry out the same function in their gut.

2–2 FATS

Fat molecules are also composed solely of carbon, hydrogen, and oxygen atoms. In contrast to carbohydrates, however, the ratio of hydrogen atoms to oxygen atoms is a great deal higher than 2:1. Tristearin, to take a common example of a fat, has the molecular formula $C_{57}H_{110}O_6$.

The high proportion of hydrogen in a fat molecule tells us that the molecule is much less oxidized than a carbohydrate molecule. This means that a given weight of fat stores a good deal more energy than the same weight of carbohydrate. Pound for pound, fats provide the most concentrated energy reserve available to the organism. Figure 2–6 shows a Brazil nut burning like a candle, thanks to its rich oil stores.

Fig. 2–6

The nut is supported by three pins.

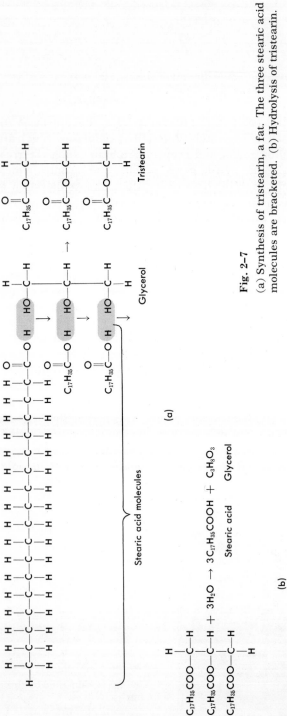

(a)

(b)

$$C_{17}H_{35}COO-C-H$$

$$C_{17}H_{35}COO-C-H + 3H_2O \rightarrow 3C_{17}H_{35}COOH + C_3H_8O_3$$

$$C_{17}H_{35}COO-C-H$$

Stearic acid   Glycerol

Fig. 2-7

(a) Synthesis of tristearin, a fat. The three stearic acid molecules are bracketed. (b) Hydrolysis of tristearin.

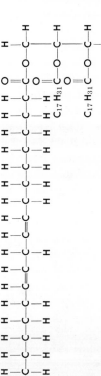

Fig. 2-8

Structure of trilinolein, an unsaturated fat. There are two double bonds in each of the fatty acid units. Unsaturated fats liquefy at lower temperatures than do saturated fats and hence are called oils.

The fat molecule is made up of four parts: a **glycerol** molecule and three molecules of **fatty acids.** Each of the three —OH groups on the glycerol molecule is capable of reacting with the

$$\begin{array}{c} \text{O} \\ \parallel \\ -\text{C}-\text{OH} \end{array}$$

group of a fatty acid molecule. A molecule of water is removed in this process, too, and the fatty acid becomes attached to the glycerol molecule (Fig. 2–7). The three fatty acids in a single fat molecule may be all alike or they may be different. They may contain as few as four carbon atoms or as many as 24. Curiously enough, the number of carbon atoms in the fatty acid is almost always an even number. 16-carbon (palmitic acid) and 18-carbon (stearic acid) fatty acid molecules are the most common in fats.

Just as polysaccharides can be hydrolyzed into their constituent sugar units, so fats can be hydrolyzed into glycerol and fatty acids (Fig. 2–7). Water is necessary, of course, as well as an enzyme called a **lipase.** The hydrolysis of fat is an important step in the digestive process.

Some fatty acids have one or more double bonds in the molecule (Fig. 2–8). Fats formed from these molecules are known as unsaturated fats. As they are likely to melt at temperatures below room temperature, we call them oils. Cottonseed oil and olive oil are two examples. These oils should not be confused with oils derived from petroleum. Their chemical nature is quite different.

In general, animal fats are saturated; plant fats unsaturated. There is a growing body of evidence that a diet rich in saturated fats predisposes one to the development of diseases of the heart and circulatory system. Unsaturated fats, on the other hand, seem not to have this effect.

Fats are insoluble in water and before they can be transported in the blood, they must be converted into a form that will mix with water. One way in which this is done is to substitute a phosphorus-containing molecule for one of the three fatty acid molecules. The resulting substance, a **phospholipid,** mixes with water and thus can be transported in the watery environment of the body. Phospholipids also play an important structural role in the organism. In close association with protein, they participate in forming the membranes from which many cell structures are built.

There is another way in which fats can be converted into a form which will mix with water. This is to emulsify them. **Emulsions** are colloids containing colloidal size droplets of one liquid suspended in another. Homogenized milk is an example of an emulsion of fat (cream) in the watery portion ("skim milk") of the milk. In order to keep the droplets of an emulsion from meeting and combining, it is necessary to coat them with a third substance, an emulsifying agent. Soaps and synthetic detergents are effective emulsifying agents for fats. We use these substances to emulsify fat deposits on our bodies and dishes. In the form of an emulsion, the fat can then be easily rinsed away.

A good deal of the fat we eat is first emulsified in the digestive tract before it is used by the body. The body's emulsifying agent is a fluid called bile which is secreted into the intestine by the liver. The active ingredients in bile are substances called bile salts. They are members of a large group of closely related chemical compounds, the **steroids.** All steroids contain a skeleton of seventeen carbon atoms organized into four rings (Fig. 2–9). Each different steroid has its own side groups attached to several points on the skeleton. The most abundant steroid in the human body is cholesterol (about one-half pound per person).

Fig. 2–9

Progesterone, an important sex hormone in women. Like all steroids, it is constructed from a skeleton of 17 carbon atoms (shown in color). Each short dash represents a hydrogen atom.

From it are synthesized the bile salts, one of our vitamins, and a number of important hormones. All the sex hormones, of which progesterone is one (Fig. 2–9), are steroids.

2–3 PROTEINS

Proteins are among the most complex of all the organic compounds. They are made up of macromolecules which contain carbon, hydrogen, oxygen, and nitrogen atoms. Sulfur atoms are usually present, too. Certain proteins contain phosphorus or some trace metal element, such as iron or copper, in addition to the other elements.

Most protein molecules are so large and complex that it has not yet been possible to learn much about their structure except their size and, in some cases, their molecular formula. Beta-lactoglobulin, a protein with a relatively small molecule, has the molecular formula $C_{1864}H_{3012}O_{576}N_{468}S_{21}$. Its molecular weight is about 42,000. Many proteins are even larger than this. Hemocyanin, an oxygen-carrying pigment found in some crustaceans, has a molecular weight greater than six million.

At first glance, it might seem a hopeless task to learn more about the structure of such enormous molecules. Fortunately, the problem is not quite so insurmountable as it might seem. Despite their great size, protein molecules are built up in an orderly way. They are polymers consisting of long chains of relatively simple monomers called amino acids. There are twenty different kinds of

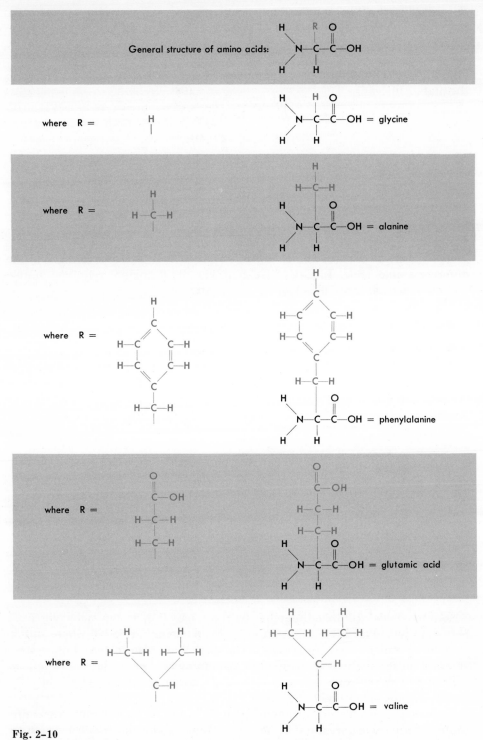

Fig. 2–10

Structures of five common amino acids. There are some 15 others generally found in proteins.

amino acids commonly found in proteins. Each of these is built according to the plan:

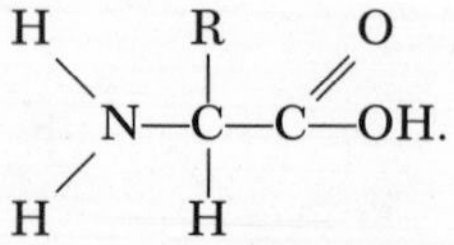

$$\begin{array}{c} H \quad\quad R \quad\quad O \\ \diagdown \quad\quad | \quad\quad \diagup\diagup \\ N-C-C-OH. \\ \diagup \quad\quad | \\ H \quad\quad H \end{array}$$

The —NH$_2$ group is called the amino group. The

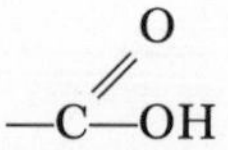

$$\begin{array}{c} O \\ \diagup\diagup \\ -C-OH \end{array}$$

group is the acid group. The chemical nature of R is different for each of the different amino acids. In every case, however, the R group is relatively simple and its structural formula is known (Fig. 2–10).

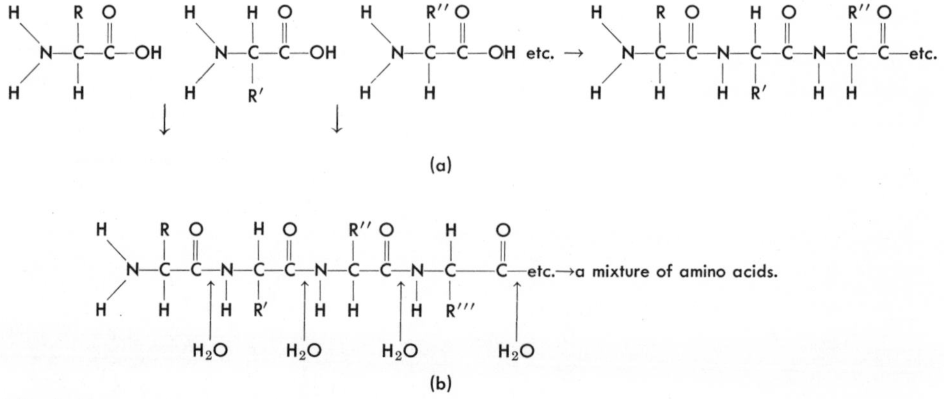

(a)

(b)

Fig. 2–11

(a) Synthesis of protein. (b) Hydrolysis of protein.

Amino acids are linked together by the removal of water molecules from between them (Fig. 2–11). Crosslinks between chains are found where sulfur-containing amino acids are present. The chains are also folded. Figure 2–12 shows a model of the arrangement of the amino acid chain in the myoglobin molecule. This molecule is one-quarter of the size of the hemoglobin molecule, having a molecular weight of 17,000.

Although even the most complex proteins are made up of simple, repeating units of amino acids, we are still a long way from working out detailed structural formulas for them. In fact, it was not until 1954 that the structural formula of the first protein was established. The protein was insulin, one of the most im-

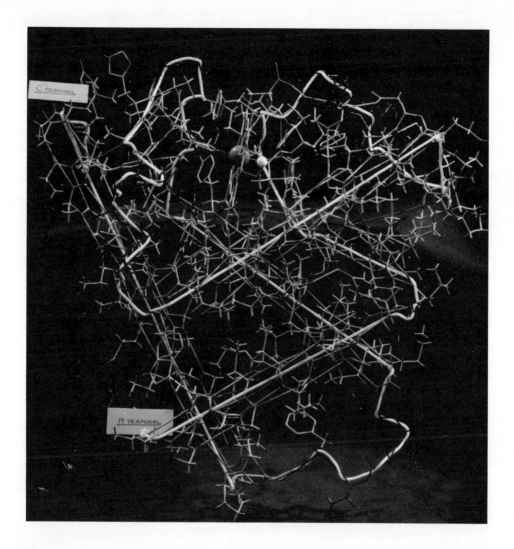

Fig. 2–12

Three-dimensional model of the myoglobin molecule. About 150 amino acids are present in the single, folded chain. The dark sphere represents the iron atom at the center of the prosthetic group, heme. The light sphere represents a molecule of water. (Courtesy of Dr. John C. Kendrew.)

portant hormones in our body. Its molecular formula, $C_{254}H_{377}N_{65}O_{75}S_{6}$, shows that its molecule is small compared with most proteins. Nevertheless, it took Dr. Frederick Sanger and his colleagues at the University of Cambridge in England ten years (1944–1954) to work out the sequence of amino acids and the locations of the crosslinks in this protein. Using clever, but tedious, steps of

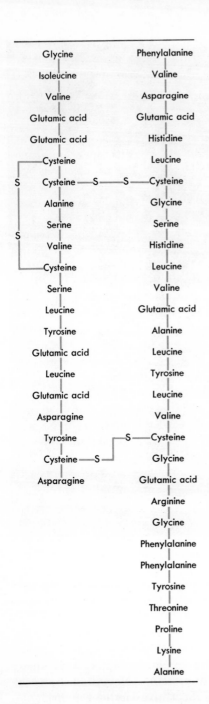

Fig. 2–13

Sequence of amino acids in the insulin molecule. The insulin molecule consists of two polypeptide chains held together by sulfur bridges.

chemical degradation, they determined the exact sequence of the 51 amino acids present in the molecule (Fig. 2–13). Since their discovery, other workers have applied similar techniques in an effort to work out the structure of still other proteins.

Knowledge of the exact sequence of amino acids in a protein naturally raises the possibility of synthesizing the protein in the "test tube." The task of assembling dozens of amino acids in a precise sequence is a formidable one, but it has been accomplished for insulin. And, in 1969, the enzyme ribonuclease, which is also a protein, was successfully synthesized. There are 124 amino acids (in a single chain) in the ribonuclease molecule.

Proteins, like starches and fats, are degraded by hydrolysis. The peptide linkages are opened with the insertion of a molecule of water at that point (Fig. 2–11). This breaks the protein molecule into shorter chains of amino acids called **polypeptides.** Finally the polypeptides are completely hydrolyzed into their constituent amino acids.

Life as we know it could not exist without proteins. They are the chief molecules out of which living cells are made. Some protein molecules are dissolved or suspended in the watery contents of the cell. Others are incorporated into various structures of the cell.

Many cell proteins are found chemically united with other kinds of molecules. These proteins are called **conjugated proteins.** Protein-lipid, protein-nucleic acid and protein-pigment combinations all play important roles in living cells. All enzymes are proteins and many of them are conjugated with smaller molecules **(prosthetic groups)** such as metal-containing pigments and vitamins.

Proteins are also found outside of cells. These extracellular proteins are important supporting and strengthening materials in animals. (Plants differ in this respect from animals. Their cells are supported and strengthened by what extracellular, organic molecules?) Bone, cartilage, tendons, and ligaments are all examples of supporting structures which contain substantial quantities of extracellular proteins, such as collagen.

Although we can make statements about the general kinds of proteins found in different organisms, it is important to realize that these proteins may not be exactly alike. Every species manufactures proteins unique to that species. Furthermore, even the individuals within a species may have some protein molecules that are absolutely unique. With the exception of identical twins, we are quite sure that no two human beings, for example, have precisely the same proteins.

At first glance, it seems unbelievable that each individual now living on earth has certain protein molecules unique to him. When you appreciate the great complexity and possibilities for variety in protein molecules, however, this specificity seems more reasonable. Just think for a moment of the large number of words in an English unabridged dictionary. Every one of these is made up from an alphabet of 26 letters. Proteins are made up from an "alphabet" of about twenty amino acids. Then think of the dictionaries in other languages that also use the Roman alphabet. Consider, too, that it is a rare word indeed that contains more than twenty letters. A single protein molecule, on the other hand, may contain thousands of amino acids.

The amino acids in proteins are not arranged in just one dimension as are the letters in a word and as they are shown in Fig. 2–13. Instead, the chains of amino acids are folded into complex, three-dimensional shapes (Fig. 2–12). This folding occurs because of interaction between certain amino acids in the chain. For example, the sulfur (—S—S—) bridges that form between the cysteine units in the chain (Fig. 2–13) bring these links in the chain close together.

The importance of the three-dimensional structure of proteins is nicely illustrated by the phenomenon of **denaturation.** Proteins are quite sensitive to a variety of chemical and physical agents. When exposed to these agents, they may lose their characteristic pattern of folding. This is denaturation. It can occur under conditions that do not affect the actual peptide linkages themselves. Nevertheless, the biochemical properties of the protein may be completely altered. For example, denaturation destroys the activity of enzyme molecules. When the white of an egg cooks, the protein (albumin) is undergoing denaturation. When more gentle methods of denaturation are used, the process may be reversible. In such cases, removal of the denaturing agent allows the protein to regain its original three-dimensional shape and full chemical activity.

Proteins can also be used by living organisms as a source of energy. If more protein is consumed in the diet than is needed to satisfy the structural needs of the body (growth and repair), the excess can be used as fuel. After the protein is hydrolyzed into its constituent amino acids, the nitrogen-containing, or amino, portion of these molecules is removed. In man, this process, called **deamination,** occurs chiefly in the liver. The non-nitrogen-containing residue is then oxidized in the same way as carbohydrate and fat materials.

2–4 NUCLEIC ACIDS

Nucleic acid molecules are even larger than protein molecules, with molecular weights that run into the billions. They, too, are polymers. The monomers of nucleic acid molecules are called **nucleotides.** Each consists of three subunits: a sugar, a phosphate, and a nitrogenous base. The sugar and phosphate groups alternate as links in the chain. The nitrogenous bases project out from the axis of the chain, with one base attached to each sugar link (Fig. 2–14).

There are two distinct kinds of nucleic acid. One is called deoxyribonucleic acid or DNA. The other is ribonucleic acid or RNA. DNA is found localized within the control center of the cell, the nucleus. Some RNA is also found there, but more of it is found in the surrounding cell contents, the cytoplasm.

RNA gets its name from its sugar unit, a five-carbon sugar called ribose. Attached to each ribose unit is one of four possible nitrogenous bases, cytosine, uracil, adenine, and guanine (C, U, A, and G). The first two of these bases are

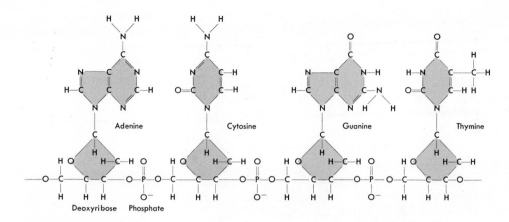

Fig. 2–14

Molecular organization of DNA. In RNA the pyrimidine uracil is found in place of thymine.

single-ring substances called pyrimidines. The second two are double-ring substances called purines.

The sugar unit in DNA is deoxyribose. As its name suggests, it has one less oxygen atom than ribose. The pyrimidines, cytosine and thymine, and the purines, adenine and guanine (C, T, A, and G), are the bases found in DNA.

Normally, DNA differs from RNA in another important respect. Its molecule is made up of not one, but two chains of alternating sugars and phosphates. These two chains are twisted around each other something like a double spiral staircase (Fig. 2–15). At each sugar, a purine or pyrimidine projects into the "stairwell." Here it joins with the pyrimidine or purine projecting out from the opposite staircase. Weak forces of attraction between these bases are sufficient to hold the two chains together.

The most interesting thing about this structure is that there is not room enough across the "stairwell" for two purines to fit at the same level, and there is too much room for two pyrimidines to meet and bond together. Consequently, a purine on one chain can be paired only with a pyrimidine on the other. Furthermore, the location of the bonds is such that the purine adenine can pair only with thymine. Similarly, guanine always pairs with cytosine. Thus the two interlocking strands of a DNA molecule are complementary to each other. In a real sense, one is the "negative" of the other. If we know the sequence of bases on one strand, we know the sequence of bases on the other.

In recent years many biologists have turned their attention to the nature and role of DNA and the three types of RNA that are found within cells. Their interest stems from an ever-growing conviction that within these molecules resides the information that controls the structure and activities of all organisms and the machinery for translating this information. They see DNA as a master

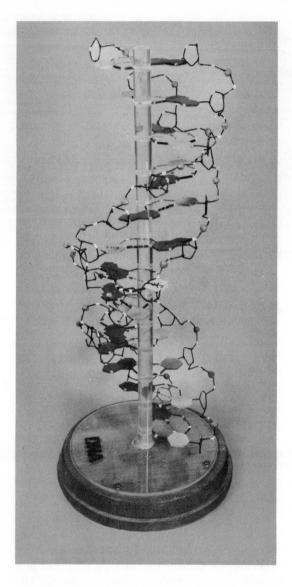

Fig. 2-15

The Watson-Crick model of DNA structure. The small spheres represent phosphate groups; the open pentagons represent deoxyribose. The solid planar structures represent the purine and pyrimidine bases. (Photo courtesy Dr. Donald M. Reynolds.)

blueprint, written in the code of purine-pyrimidine pairs, for the building and operating of a living organism. They see each of the three major types of RNA as playing an essential role in the execution of the instructions coded within the DNA molecules. (Some of the discoveries that have led to these ideas will be discussed in the next chapter and in Chapter 7.)

Perhaps one day the fundamental nature of life itself can be understood in molecular terms; it begins to look as though the study of the nucleic acids will bring us closer to this goal than ever before.

Fig. 2–16

Some of the principal vitamins.

VITAMIN	DEFICIENCY DISEASE	SOURCES	OTHER INFORMATION
A	Night blindness	Milk, butter, fish liver oils, carrots, and other vegetables	Precursor in the synthesis of the light-absorbing pigments of the eye Stored in the liver Toxic in large doses
Thiamine (B$_1$)	Beriberi Damage to nerves and heart	Yeast, meat, unpolished cereal grains	Coenzyme in cellular respiration
Riboflavin (B$_2$)	Inflammation of the tongue Damage to the eyes General weakness	Liver, eggs, cheese, milk	Prosthetic group of flavoprotein enzymes used in cellular respiration
Nicotinic acid (Niacin)	Pellagra (Damage to skin, lining of intestine and perhaps nerves)	Meat, yeast, milk	Converted into nicotinamide, a precursor of NAD and NADP—two important coenzymes for REDOX reactions in the cell
Folic acid	Anemia	Green leafy vegetables Synthesized by intestinal bacteria	Used in synthesis of coenzymes of nucleic acid metabolism
B$_{12}$	Pernicious anemia	Liver	Each molecule contains one atom of cobalt.
Ascorbic acid (C)	Scurvy	Citrus fruits, tomatoes, green peppers	May act as a reducing agent in the body
D	Rickets (Abnormal Ca^{++} and PO$_4^{\equiv}$ metabolism resulting in abnormal bone and tooth development)	Fish liver oils, butter, steroid-containing foods irradiated with ultraviolet light	Synthesized in the human skin upon exposure to ultraviolet light Toxic in large doses
E	No deficiency disease known in humans	Egg yolk, salad greens, vegetable oils	–
K	Slow clotting of the blood	Spinach and other green leafy vegetables Synthesized by intestinal bacteria	Necessary for the synthesis of prothrombin, an essential agent in the clotting of blood

2–5 VITAMINS

The carbohydrates, fats, proteins, and nucleic acids are the most prevalent organic molecules found in living things. There are others such as organic acids, alcohols, and steroids, that play important roles, too. These are all small molecules, however, and do not contribute to the fixed, structural components of the cell. Furthermore, almost all of them can be synthesized by the organism from one or more of the major kinds of molecules that we have been discussing.

The few organic molecules that an organism cannot manufacture from carbohydrate, fat, protein, or nucleic acid sources are called vitamins. With a few exceptions, green plants have no vitamin requirements as they manufacture every organic molecule they need. Among these, however, are substances that are vitamins for nongreen plants, microorganisms, and animals. It is important to realize that the term vitamin does not refer to a particular chemical group. The only way in which the various vitamins are similar to one another is that (1) they are all organic, (2) they are not used as an energy source or for construction of the cell, and (3) the organism cannot synthesize them from the standard foods in its diet. Note, too, that what may be a vitamin for one organism is not necessarily a vitamin for another.

Vitamins are needed in only very small quantities. Years of research have shown that some of them, at least, are incorporated into enzymes. These vitamins are conjugated as a "prosthetic group" with the protein part of the enzyme. In view of this role, it is not surprising that their presence is vital to the welfare of the organism. If it cannot synthesize them, then it must include them in its diet. If it fails to do so, a **deficiency disease** may result (Fig. 2–16).

2-6 MINERALS

Although living organisms are distinguished by the complex array of organic molecules of which they are made, inorganic substances play an important part, too. Aside from C, H, O, N, and S, most of the elements found in living things are present in the form of ions of salts. These are the inorganic minerals of the body. Some of these minerals are quite insoluble in water and form solid deposits. Many organisms exploit these as supporting and protecting structures. Calcium carbonate ($CaCO_3$) is the principal constituent of the shells of mollusks. The bones of vertebrate skeletons contain calcium carbonate, large amounts of calcium phosphate [$Ca_3(PO_4)_2$] and some magnesium and fluoride ions. All these substances provide rigid support and protection for soft tissues of the body. Inorganic ions also carry out other important roles in living things. We will examine some of these in Chapter 4.

2-7 WATER

Water is also an inorganic substance and an indispensable constituent of life. Figure 2–1 shows that 66% of the weight of the human body is water. The water content of most cells is about the same, although some of the jellyfishes are more than 90% water.

Water has several important properties that make it an ideal constituent of living things. One of these is that it remains liquid over the range of temperatures generally found on earth. In liquid form it is an excellent solvent for thousands of other substances, both organic and inorganic. Its liquid nature and great solvent power make it unexcelled as a transport medium. Blood, which is

90% water, serves to transport materials throughout our body. The fact that water dissolves so many substances means that these substances can be brought together in the form of individual molecules and ions. In this form, reactions between different substances can occur quickly. Water thus serves as the medium in which almost every chemical reaction in living things takes place. Water is also a reactant and/or product of many chemical reactions in living organisms. (Can you name one?)

Another important property of water is the slowness with which it changes temperature or changes from solid (ice) to liquid, or liquid to gas, as heat is added. It takes more heat to accomplish these changes with water than with almost any other substance known. (Our unit of heat, the calorie, is, in fact, defined as the quantity of heat needed to raise one gram of water one degree Celsius—see table on page 194.) This property of water is extremely important to life. It means that the temperature of living organisms changes relatively slowly despite sudden temperature changes in the environment. The great amount of heat absorbed when water vaporizes is exploited in sweating. Evaporation of water from the skin absorbs heat from the body, thus cooling it. Not only do the thermal properties of water affect organisms directly, but they also bring about a moderating effect in the environment. Freshwater and marine organisms are subject to a far smaller range of temperatures than land-living animals of the same region. Even land-living animals benefit from the temperature-moderating effect of nearby bodies of water. The ocean's slow absorption of heat in the summer and slow release of heat in the winter tend to reduce violent extremes of temperature in coastal regions. It is no accident that record low temperatures as well as record high temperatures are made in inland areas such as central Siberia and central North America.

Even when water does get cold enough to freeze, the ice that is formed is less dense than the water, and thus floats on top of the water. This enables aquatic organisms to remain surrounded by water under the ice and also provides for speedy melting of the ice in the spring. The solid form of most substances is more dense than the liquid. It is interesting to speculate on what life in the "temperate" regions of the earth would be like if water were not an unusual substance in this respect.

In this chapter we have examined some of the chemical substances of which living things are composed. In the next chapter, we will try to learn how these chemical substances are organized to make up the various structures of the basic unit of life itself, the cell.

EXERCISES AND PROBLEMS

1 Pound for pound, what food yields the most metabolic water when oxidized?

2 What element is found in all proteins but in no oils?

3 Summarize the importance of water to living things.

4 What is a vitamin?

5 How does RNA differ from DNA?

6 Write the structural formula for the neutral fat tripalmitin (palmitic acid = $C_{15}H_{31}COOH$).

7 Write the structural formula for the tripeptide containing glycine—alanine—phenylalanine.

8 Show by means of structural formulas the hydrolysis of the tripeptide described in Question 7.

9 Distinguish between saturated and unsaturated fats.

10 Name an isomer of glucose; of sucrose.

REFERENCES

1 Doty, P., "Proteins," *Scientific American,* Reprint No. 7, September, 1957. Their chemical and physical make-up.

2 Thompson, E. O. P., "The Insulin Molecule," *Scientific American,* Reprint No. 42, May, 1955. Describes the techniques used to determine the *chemical* structure of the insulin molecule.

3 Kendrew, J. C., "The Three-Dimensional Structure of a Protein Molecule," *Scientific American,* Reprint No. 121, December, 1961. Describes the techniques used to determine the *physical* structure of the myoglobin molecule.

4 Crick, F. H. C., "Nucleic Acids," *Scientific American,* Reprint No. 54, September, 1957. The author shared a Nobel prize for his work in deducing the organization of nucleic acid molecules.

Cells of the blue-green alga *Plectonema boryanum*. While the cells of this organism do not have many of the structures described in this chapter, they are nonetheless exceedingly intricate in their organization and capable of carrying out all the functions of life. (41,000 X, courtesy Kenneth M. Smith and R. Malcolm Brown, Jr.) ▶

CHAPTER 3   CELL STRUCTURE
AND FUNCTION

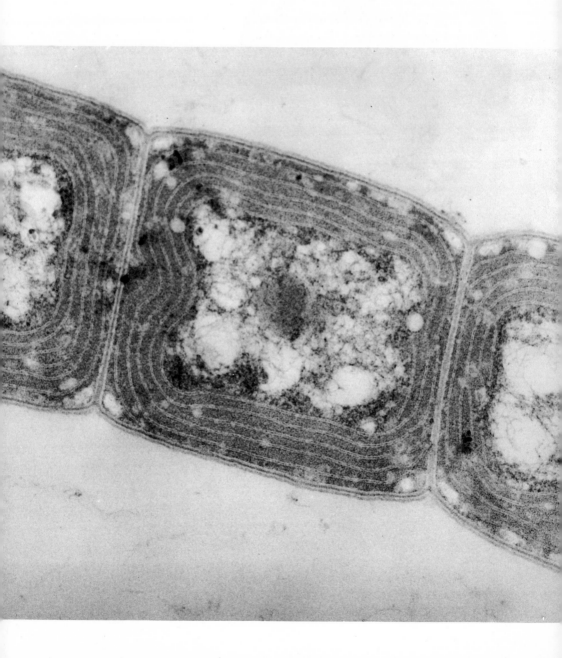

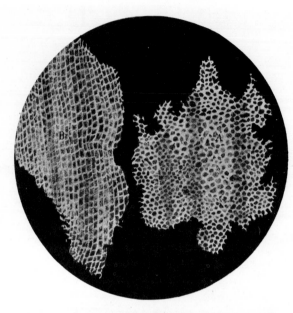

Fig. 3-1
Robert Hooke's drawings of the
cellular structure of cork were
published in 1665. (Courtesy
Bettmann Archive.)

The carbohydrate, lipid, protein, nucleic acid, and other molecules that make up
living things are not themselves alive. It is only when these molecules become
organized in precise ways that the phenomenon of life appears. The minimum
organization of this matter that can live is called the cell.

3-1 THE CELL "THEORY"

We owe the term *cell* to the research of the Englishman, Robert Hooke. In 1665,
he told of examining thin slivers of cork under his microscope and finding them
to be made up of many neatly arranged little chambers (Fig. 3-1). He called
these chambers *cells*. All that Hooke saw, however, were the cellulose walls
laid down by once-living material. Today we use the term *cell* to include the active, living contents of such chambers. Sometimes the term *protoplast* is used
when one wishes to emphasize that the active materials within any outer covering or wall are being considered.

The term *protoplast* is itself derived from the word protoplasm. Protoplasm
is the material that makes up the protoplast. The term is falling into disuse,
however, because no single material does make up the protoplast. As we have
seen, living material is composed of a large variety of different molecules. There
is no single living substance. Actually, the word can still be useful to the biologist, but only as a shorter equivalent of "living material" or "the molecules that
make up living things." In this sense, one can speak of growth resulting in an
increase in the amount of protoplasm. On the other hand, one should not use
the term as though it referred to a single, specific substance. To say that "living

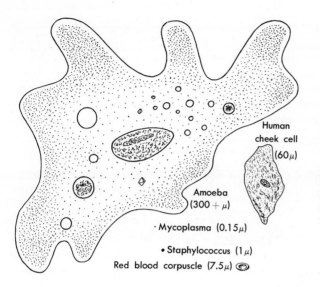

Human
cheek cell
(60μ)

Amoeba
(300 + μ)

· Mycoplasma (0.15μ)

• Staphylococcus (1μ)

Red blood corpuscle (7.5μ)

Fig. 3–2

Comparative sizes of various cells. The amoeba is just visible to the unaided eye. Mycoplasmas can be seen clearly only under the electron microscope.

things are made of protoplasm" in the sense that one says "sand is made of silica (SiO_2)" is as meaningless as to say that "television sets are made of videoplasm."

The idea that the cell is the fundamental unit of which all living things are constructed grew out of the work of the German botanist Schleiden and the German zoologist Schwann. Schleiden trained his microscope on a wide variety of plant structures and in every case found that they were made of cells. Schwann made the same discovery in his microscopic examination of animal parts. Their dual discovery, which was set forth in 1839, is called the cell theory. It is not really a theory, though, but an observable fact of nature.

3–2 SIZES AND SHAPES

Cells occur in a variety of sizes and shapes. Eggs are single cells and most of them can be seen easily with the unaided eye. This is because of the bulky food reserves they contain. Some of the large unicellular microorganisms are just visible to the naked eye. An amoeba cell, some 300μ across, is about the size of the periods printed on this page. Tissue cells are generally microscopic. In man, they may be as small as the red blood corpuscles, which are 7.5μ in diameter, or as large as the cells lining the inner surface of the cheek, which are some ten times that size. The smallest bacteria are just under 1μ in diameter, while the smallest microorganisms known, the mycoplasmas, are only a tenth of that. The mycoplasmas probably represent the smallest size a cell can be and still have all the attributes of life (Fig. 3–2).

Cells vary markedly in shape as well as size. The shape of a cell is usually a clear reflection of the function it carries out in the organism. Nerve cells, which must transmit electrical impulses over long distances, have long extensions, in some cases several feet in length. Muscle cells are elongated so that the force of contraction can be exerted efficiently in one direction. The supporting cells of a plant have thick walls. Where masses of cells are simply packed together, they assume the shape of a polyhedron, a solid with many faces. This shape permits close packing of the cells without a large amount of space between them.

3-3 THE CELL MEMBRANE

The outer surface of all protoplasts is bounded by a very thin, elastic membrane. It is so thin (75–100 A*) that its structure can be observed only under the electron microscope.

  The cell membrane provides an important barrier between the interior of the cell and the exterior. Water and many other small molecules pass through it quite easily. On the other hand, many ions such as the sodium ion (Na^+) and large molecules such as protein molecules pass through with greater difficulty. A membrane that permits the free passage of some materials and not others is called a semi- or differentially permeable membrane. The permeability of the cell membrane changes with changing conditions and this suggests that it is no passive barrier but exerts a selective action on which materials pass through it.

  Chemical and physical studies of the cell membrane suggest that it is made up of three layers, of roughly equal thickness, sandwiched together. The outer layer seems to be composed chiefly of protein molecules. The middle layer is actually a double layer of phospholipid molecules (see Section 2–2). The inner layer is another protein layer. The protein layers account for the elasticity of the membrane, since the folded protein molecules that make them up can be stretched a moderate amount. The lipid layer accounts for the ease with which many fat-soluble molecules pass through the cell membrane.

  The picture of the cell membrane that has been derived from chemical and physical studies has been nicely supported by electron micrographs. Figure 3–3 shows the membranes of two adjacent cells. The sandwich construction of the membranes is clearly visible.

  Membranes of this construction bound the cells of all organisms: microorganisms, plants, and animals. Membranes of similar, often identical, construction are also found within the interior of cells. Here they serve to establish a variety of special compartments. In fact, the protein-lipid-protein membrane seems to be such a fundamental structural unit throughout the cell that it has been given a special name: the **unit membrane.**

  Except in the bacteria and blue-green algae, there are two major compartments of the cell, the cytoplasm and the nucleus.

---

* One angstrom (A) = $10^{-4}\mu$ = 0.1 mμ.

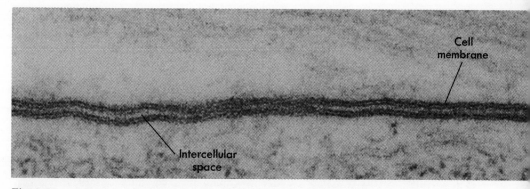

Fig. 3-3

Cell membranes (140,000 X). The membranes of two adjoining cells are shown here. The triple-layer construction of each of these "unit" membranes shows clearly. (From Fawcett: *The Cell: Its Organelles and Inclusions,* W. B. Saunders Co., 1966.)

3-4 THE NUCLEUS

The nucleus is separated from the cytoplasm by a pair of unit membranes. The envelope thus formed is not continuous but, as can be seen in Fig. 3-4, contains pores. These probably permit materials to pass between the nucleus and the cytoplasm.

Within the nuclear membrane there is a semifluid medium in which are suspended the **chromosomes.** Usually these are present as very elongated structures and cannot be easily observed under the light microscope. The term *chromatin* is used to describe them when they are in this condition.

When a cell is preparing to divide into two cells, the appearance of the chromosomes changes. The long thin strands coil up into thickened, dense bodies, which (with the help of an appropriate stain) are easily visible in the light microscope (Fig. 3-5). During the process of cell division, the chromosomes are distributed in precisely equal numbers to the two daughter cells.

Chemically, the chromosomes are made up of DNA and protein. We do not yet know whether a single chromosome contains only a single DNA molecule or many of them. The protein is a special type called *histone.*

During the period between cell divisions, when the chromosomes are in the extended state, one or more of them may have a large, spherical body attached. This body, the **nucleolus,** is easily visible in the light microscope. Here are synthesized several kinds of RNA and protein molecules (including histones). Some of the RNA manufactured here is assembled into **ribosomes.** These tiny bodies are essential for protein synthesis both here and elsewhere in the cell.

The nucleus functions as the control center of the cell. Our appreciation of its role has come from many experiments and observations. In particular, the development of micromanipulators has enabled cytologists to remove or trans-

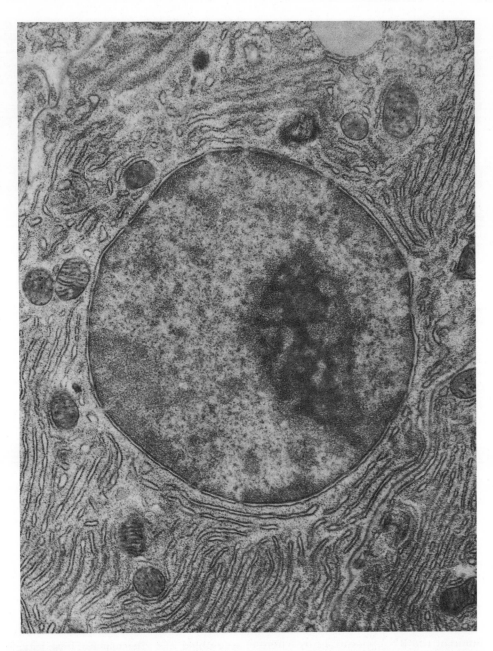

Fig. 3-4

The nucleus and surrounding cytoplasm of a cell from the pancreas of a bat (18,000 X). Note the double-layer construction of the nuclear membrane and the pores in it. What other cell structures can you identify? (Electron micrograph courtesy of Dr. Don W. Fawcett.)

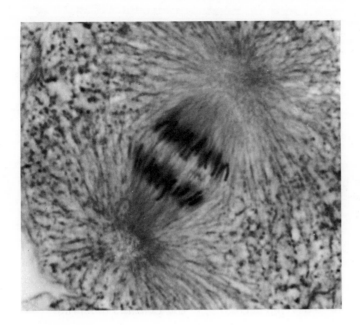

Fig. 3–5

Dividing cell in the embryo of a whitefish. The dark horseshoe-shaped structures are chromosomes. The array of microtubules to which they are attached is called the spindle. (Courtesy General Biological Supply House, Inc.)

plant nuclei from single cells and to study the results of these operations. If the nucleus is removed from an amoeba, the organism continues to live for a few days. It cannot reproduce, however, and eventually it dies. That it is the loss of the nucleus which causes this and not just mechanical damage from the operation can be shown by piercing an amoeba with the microtools but not actually removing the nucleus. Such an amoeba recovers fully from this experimental procedure.

Even before the invention of micromanipulators, the importance of the nucleus in determining what the cytoplasm does had been demonstrated by the German biologist Theodor Boveri. By vigorous shaking, Boveri was able to remove the nucleus from the eggs of a species of sea urchin of the genus *Sphaerechinus*. He then allowed these "enucleated" eggs to be fertilized by sperm from sea urchins of the genus *Echinus*. Sperm cells are much smaller than egg cells. They consist of little more than a nucleus and a tail to propel it. In the process of fertilization, the nucleus penetrates the egg. Thus fertilization of enucleated *Sphaerechinus* eggs by *Echinus* sperm resulted in the substitution of one kind of nucleus for another. The stimulus of fertilization caused the egg to undergo cell division and it developed into a sea urchin larva. A glance at

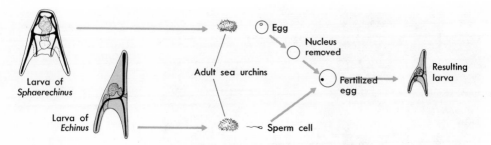

Fig. 3-6

Boveri's experiment, which demonstrated the importance of the nucleus in controlling the cell. An *Echinus* nucleus in *Sphaerechinus* cytoplasm resulted in a larva with *Echinus* characteristics.

Fig. 3-6 shows that this larva possessed the traits of the *Echinus* species rather than of the *Sphaerechinus* species. Although *Echinus* contributed little more than a single tiny nucleus to the system, this nucleus caused the great mass of *Sphaerechinus* cytoplasm to develop according to the *Echinus* blueprint.

3-5 THE CYTOPLASM

The term *cytoplasm* suffers from much the same defect as the term *protoplasm*. Cytoplasm is a complex, heterogeneous material. Basically, it consists of a rather watery medium, the "ground substance," in which is suspended a variety of distinct structures. The functions of the cytoplasm are, for all practical purposes, the functions of the structures suspended in it. We will now examine these.

3-6 RIBOSOMES

Among the smallest structures suspended in the cytoplasm are the ribosomes. These roughly-spherical bodies are so small (150 A) that they can be seen only under the electron microscope (Fig. 3-4). They are composed of RNA and protein. All the proteins of the cell, including all enzymes, are synthesized by the ribosomes. As we shall see in Chapter 8, the ribosomes literally carry out the instructions contained within the DNA code of the nucleus.

The ribosomes that are engaged in synthesizing proteins for use within the cell are distributed randomly in the ground substance of the cell. However, many cells, such as those of the liver and pancreas, also synthesize proteins that are secreted *outside* the cell. The ribosomes that synthesize these proteins are found attached to the membranes of the endoplasmic reticulum.

3–7 THE ENDOPLASMIC RETICULUM

The endoplasmic reticulum is an elaborate system of membranes that have the same basic structure as the other membranes we have described. They are organized in pairs and thus form totally enclosed spaces within the ground substance of the cell. If you look closely at Fig. 3–4, you can see rows of ribosomes adhering to the outer surface of the reticular membranes, that is, the surface in contact with the ground substance of the cell. The protein that is produced by these ribosomes is secreted into the cavity of the endoplasmic reticulum for eventual export to the outside.

Not all the membranes of the endoplasmic reticulum have ribosomes adhering to them. Those that do not (the "smooth" endoplasmic reticulum) may be active in the synthesis and/or accumulation of other types of molecules such as polysaccharides and steroids.

Prominent in cells that secrete large quantities of material is the **Golgi complex.** This consists of neat stacks of membranes like those of the endoplasmic reticulum but without any ribosomes attached to them. The spaces within the membranes of the Golgi complex may connect from time to time with the spaces of the endoplasmic reticulum. In this way proteins are transferred to the Golgi complex and assembled in granules prior to being secreted from the cell. Polysaccharides such as cellulose (in plant cells) and mucus may also be present in the Golgi complex before being transported to the outside. There may well be other important functions of this structure that have yet to be discovered.

3–8 THE MITOCHONDRIA

The mitochondria are spherical or rod-shaped bodies that range in size from 0.2μ to 5μ. The number in a cell varies, but active cells (e.g., liver cells) may have over a thousand of them.

Although the larger mitochondria can be seen under the light microscope, only the electron microscope can reveal their basic structure. Electron micrographs show that each mitochondrion is bounded by a double membrane. The outer membrane provides a smooth, uninterrupted boundary to the mitochondrion. The inner membrane is repeatedly extended into folds that project into the interior space of the mitochondrion (Fig. 3–7). These shelflike inner folds are called **cristae.**

The membranes of the mitochondria appear to be similar to the cell membrane. Like it, they consist of a double layer of phospholipid molecules sandwiched between layers of protein molecules. Also like cell membranes, they can expand and contract. In fact, mitochondria are frequently observed to enlarge or shrink as the metabolic activity of the cell varies.

The function of the mitochondria is quite clear. They contain the enzymes that carry out the oxidation of food substances. Thus the mitochondria convert

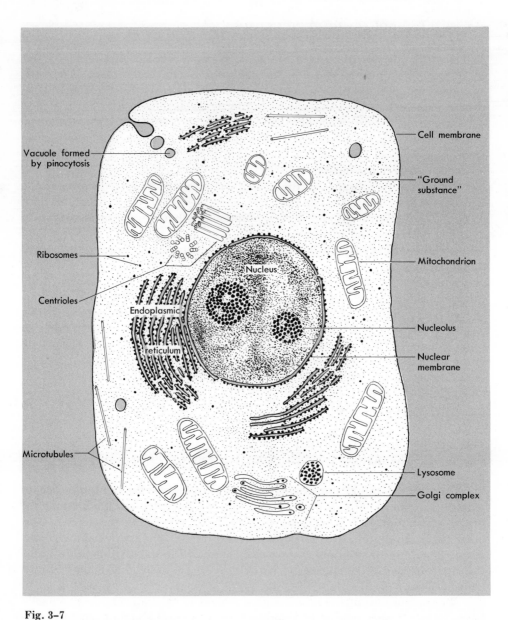

Fig. 3–7

An idealized view of an animal cell as it might be seen under the electron microscope. Although no single electron micrograph has revealed all the structures shown, the drawing represents a composite view of what many electron micrographs suggest is the organization of the parts of the cell.

the potential energy of different food materials into a form of energy that can be used by the cell to carry out its various activities. In view of this, it is not surprising that mitochondria tend to congregate in the most active regions of a cell. Nerve cells, muscle cells, and secretory cells all contain many mitochondria located in the regions of the cell most actively engaged in transmission of electrical impulses, contraction, and secretion, respectively. The mitochondria have been aptly called the powerhouses of the cell.

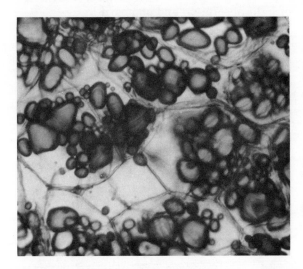

Fig. 3-8
Starch grains in the cells of a white potato. Note the cell walls. The starch grains have been lightly stained with iodine.

3-9 PLASTIDS

Plastids are found only in plant cells and the cells of most algae. Some plastids contain red, yellow, or orange pigments and impart these colors to flowers and ripe fruits. Some plastids are colorless. These are usually found in plant parts that are not exposed to the light and serve as a storage area for starch. The starch accumulates in grains. Figure 3-8 shows starch grains in the cells of the white potato.

By far the most important plastids are the chloroplasts. These contain the green pigment **chlorophyll**. It is chlorophyll which traps the energy of sunlight and enables it to be used for the manufacture of food. Thus the chloroplasts are the site at which photosynthesis takes place.

In plants, the chloroplasts are usually disk-shaped structures, 5-8μ in diameter and 1μ thick. Photosynthesizing cells may contain as many as 50 or more of them. Among the algae, the chloroplasts may assume a wide variety of shapes and only one or a few be present in a cell. Figure 3-12 shows the single, cup-shaped chloroplast in the microscopic green alga *Chlamydomonas*.

Chemical studies show that chloroplasts are rich in structural protein and phospholipids. This suggests that a protein-lipid-protein unit membrane may also be present in chloroplasts, and electron micrographs have indeed revealed its presence. Under the electron microscope, the chloroplast is seen to be bounded by a unit membrane and also to contain interconnecting stacks of membranes in its interior (Fig. 3–9). These, in turn, appear to be made up of layers of chlorophyll and lipid molecules sandwiched between the layers of protein. This arrangement spreads the chlorophyll molecules over a large area and seems to provide for maximum efficiency of light absorption and photosynthesis.

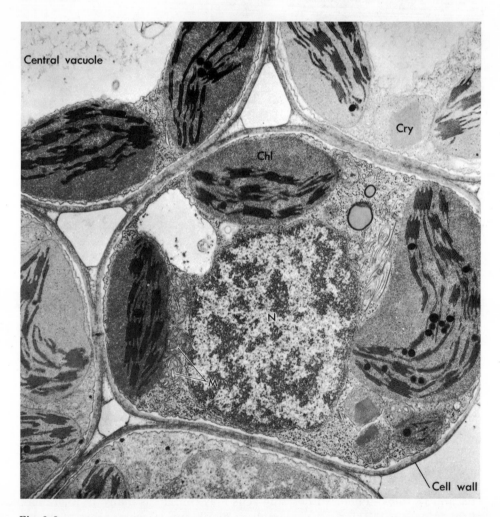

Fig. 3–9
Cells from the leaf of a sunflower. Note the nucleus (N), chloroplasts (Chl), mitochondria (M), crystals (Cry), central vacuole, and cell wall in these typical plant cells. (Electron micrograph courtesy of H. J. Arnott and Kenneth M. Smith.)

3–10 VACUOLES

Vacuoles are fluid-filled "bubbles" in the cytoplasm. They are bordered by a unit membrane that is probably identical to the cell membrane. In fact, vacuoles are often formed by an infolding and pinching-off of a portion of the cell membrane (Fig. 3–7). Food materials or wastes may be found inside vacuoles.

A young plant cell contains many small vacuoles, but as the cell matures, these unite to form a large **central vacuole** (Fig. 3–9). Dissolved food molecules, waste materials, and pigments may be found in it.

3–11 LYSOSOMES

Lysosomes (Fig. 3–10) are cell structures about the size of mitochondria and bounded by a unit membrane. Within them are contained a number of enzymes that catalyze the breakdown of polysaccharide, protein, and nucleic acid mole-

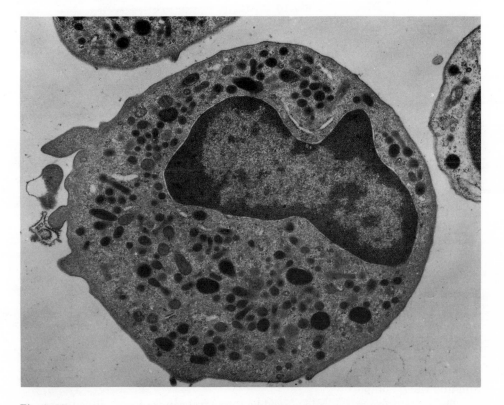

Fig. 3–10

White blood corpuscle from a guinea pig. The small dark bodies in the cytoplasm are lysosomes. (9500 X, from Fawcett: *The Cell: Its Organelles and Inclusions*, W. B. Saunders Co., 1966).

cules. Presumably, these enzymes are kept within the lysosomes to prevent breakdown of the cell itself.

When materials within the cell are to be digested, they are first incorporated into a lysosome. These materials may be other subcellular structures, such as mitochondria, that have ceased to function efficiently. They may be food particles that have been taken into the cell. In the case of the white blood cell shown in Fig. 3–10, they are bacteria or other harmful particles that have been scavenged by the cell.

Lysosomes also play an important role in the death of cells. When a cell is injured or dies, its lysosomes aid in its disintegration. This clears the area so that a healthy cell can replace the damaged one. Cell death is also a normal process in some organisms. For example, as a tadpole changes into a frog, its tail is gradually absorbed. The tail cells, which are richly supplied with lysosomes, die and the products of their disintegration are used in the growth of new cells in the developing frog.

Lysosomes have not been conclusively shown to occur in plant cells.

3–12 CRYSTALS AND OIL DROPLETS

Crystals are deposits of food or waste materials that are surrounded by a unit membrane. As the name implies, the deposits are crystalline, that is, polyhedrons with flat faces and sharp edges. Crystals are most common in plant cells (Fig. 3–9) and are usually deposits of calcium salts. These are presumably waste products of the cell's metabolism. In animal cells, the crystals are generally protein.

Oil droplets are common in both plant and animal cells. They differ from crystals in that there is usually no unit membrane separating the oil from the rest of the cytoplasm. Under the microscope, the droplets generally appear as tiny, glistening spheres. However, in the specialized fat storage cells of some plant and animal tissues, the amount of oil present may be so great as to practically fill the entire cell. Oil droplets serve as a reserve supply of fuel for the cell.

Most of the cell structures discussed so far have unit membranes as part of their construction. In general, there seem to be two main functions accomplished by these membranes. One is simply to establish a number of compartments within the cell. The bearers of the hereditary code, the chromosomes, are separated from the rest of the cell by the nuclear membrane. The potentially destructive digestive enzymes in the lysosomes are kept from contact with the ground substance of the cytoplasm by their bounding membrane. The secretory products of the cell are sequestered in the channels of the endoplasmic reticulum and Golgi complex.

The second important role played by the membranes of the cell is to provide for the neat spatial organization of enzymes and pigments. Chlorophyll is incorporated in the internal membranes of the chloroplast. Many of the enzymes which carry out the oxidation of food are neatly arranged on the cristae of the mitochondria. It is quite likely that other important enzymes are present in or on the cell membrane and the membranes of the endoplasmic reticulum.

3-13 MICROTUBULES

Microtubules are straight cylinders of protein that are found in many cells. These cylinders are about 250 A in diameter and quite long (Fig. 3-7). They are also quite stiff and therefore give some rigidity to those parts of the cell in which they are located. They often have a second function as well. In many cells, the cytoplasm (or some parts of it) flows from place to place within the cell. This is especially dramatic in the formation of pseudopodia by an amoeba (see Fig. 4-10) or a white blood corpuscle but occurs in many other cells as well. Wherever this has been observed, it appears to be associated with the presence of microtubules.

One special case of intracellular movement occurs during the precise distribution of chromosomes to the daughter cells that are formed in cell division. Each chromosome moves to its final destination attached to a long microtubule. The entire array of microtubules participating in the process is called the spindle (Fig. 3-5).

3-14 CENTRIOLES

Animal cells and the cells of some microorganisms and lower plants contain two centrioles located near the exterior surface of the nucleus. Each centriole consists of a cylindrical array of nine microtubules. However, each of the nine microtubules has two partial (as seen in cross section) tubules attached to it (Fig. 3-7). The two centrioles are usually placed at right angles to each other.

Just before a cell divides, its centrioles duplicate and one pair migrates to the opposite side of the nucleus. The spindle (see above) then forms between them.

In some cells, the centrioles duplicate to produce the basal bodies of cilia and flagella.

3-15 CILIA AND FLAGELLA

Many cells have whiplike extensions, either short ones (cilia) or long ones (flagella). In microorganisms such as *Chlamydomonas* (Fig. 3-12), cilia and flagella are used for locomotion. However, many animals have cells the cilia of which serve simply to move materials past the cell. The cells lining the inner surface of our trachea ("windpipe") are ciliated.

The origin and structure of cilia and flagella seem to be fundamentally the same. In each case they grow out of basal bodies. These have the same structure as centrioles and are formed by them.

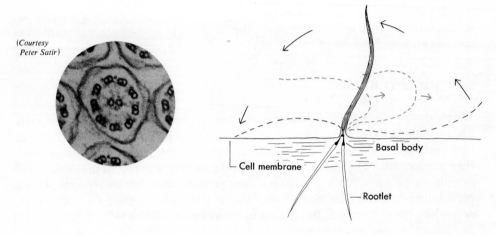

Fig. 3-11

At left, an electron micrograph of a single cilium in cross section. Note the characteristic pattern of filaments. At right, drawing of a single cilium. The power stroke is shown in black; the recovery stroke, in color.

The cilium or flagellum itself has not only the outer ring of 9 microtubules (each now with just *one* accessory tubule attached to it) but also two central fibrils that are identical to microtubules in their construction (Fig. 3-11). In both cilia and flagella, the entire assembly is sheathed in a unit membrane which is simply an extension of the cell membrane.

The similarities between the structure of cilia, flagella, basal bodies, and centrioles and the structure of microtubules suggest that the microtubule is another one of the fundamental architectural components of cells. In function, also, it is interesting that in most cases where the microtubular structure appears, it is in some way associated with cell motion.

3-16 CELL COATINGS

Only rarely is the cell membrane the outer surface of the cell. Usually some type of exterior coating is present. In animal cells this appears to be constructed from a protein-polysaccharide complex. It is not rigid but does serve to cement adjacent cells together. In Fig. 3-3, it is represented by the thin line between the two unit membranes. In many cases, however, it is much thicker than shown here. Filaments of the protein collagen (see the illustration at the start of Chapter 2) are often embedded in the thicker coatings.

In many algae and in all plants, the exterior coating is made of the polysaccharide cellulose. This forms a rigid, boxlike **cell wall** which is one of the most characteristic features of these cells. The "cells" that Hooke saw (Fig. 3-1) were the walls of once-living cork cells. In Fig. 3-9, the cell wall that surrounds each of the sunflower cells is clearly visible.

The cellulose molecules that make up the cell wall are deposited in an orderly pattern or "weave" that increases the mechanical strength of the wall. As the cell grows, the wall grows, too, in a few cases as much as a thousandfold. The cell wall is porous, permitting molecules, both small and large, to pass through it with relative ease. Unlike the cell membrane, it does not exert any control over the kinds of molecules that pass through it.

Rigid cell walls are also characteristic of bacteria and fungi. However, polysaccharides other than cellulose are used in their construction.

While not *all* cells contain *all* the structures discussed in this chapter, most of these structures are commonly found in the cells of animals, plants, and even in most microorganisms. However, two groups of microorganisms—the bacteria and blue-green algae—have cells that differ from the cells of other organisms in several notable respects. Their cells have no membrane-bounded nucleus, no mitochondria, plastids, endoplasmic reticulum, Golgi complex, or vacuoles (see illustration at the start of the chapter). The term *procaryotic* is often used to differentiate the cells of these organisms from the nucleus-containing cells of all other (*eucaryotic*) organisms. While some bacteria have flagella, these consist of simply a single strand, like a single microtubule. There is not the multi-stranded, "9 + 2" arrangement that we find in the cilia and flagella of other organisms. The absence of so many structures from the cells of organisms in these two groups suggests that these creatures are quite primitive in their organization. Nevertheless, they are capable of carrying out all the essential functions of life.

3–17 THE CELL AS THE UNIT OF STRUCTURE OF LIVING THINGS

We have seen that living things are made of one or more cells. The microscopic green alga *Chlamydomonas* is a single-celled organism. Within its single cell, it contains all the equipment needed to carry out the various functions of life. From time to time, *Chlamydomonas* divides and forms two individuals where before there was one. Each daughter cell receives a complete set of the nuclear controls present in the parent cell. Prior to actual division of the cell, each chromosome in the nucleus is duplicated. Then, during the process of cell division itself, these duplicated chromosomes become separated. With remarkable precision, a complete set migrates to each of the two daughter cells. **Mitosis** is the term used to describe this important process. It provides a mechanism for the reproduction of single-celled organisms. It also provides a mechanism for growth in multicellular organisms.

Among the flagellated green algae are several interesting colonial forms (Fig. 3–12). These species are called colonial because they are simply made up of clusters of independent cells. If a single cell of *Gonium, Pandorina,* or *Eudorina* is isolated from the rest of the colony, it will swim away looking quite like a *Chlamydomonas* cell. Then, as it undergoes mitosis, it will form a new colony with the characteristic number of cells in that colony. The situation in

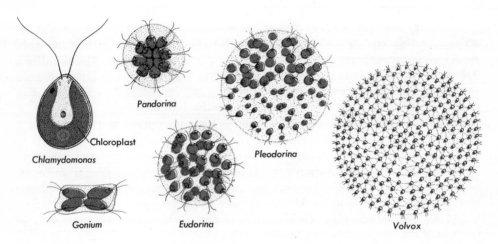

Fig. 3–12

A group of flagellated green algae whose constituent cells are strikingly similar and which illustrate the unicellular, colonial, and multicellular levels of organization. The small cells in *Pleodorina* and most of the cells in *Volvox* are incapable of reproduction. The scale of the drawings diminishes from left to right.

Pleodorina and *Volvox* is somewhat different. In these cases, some of the cells of the colony (most, in *Volvox*) are not able to live independently. If a non-reproductive cell is isolated from a *Volvox* colony, it will fail to reproduce itself by mitosis and eventually it will die. What has happened? In some way, as yet unclear, *Volvox* has crossed the line separating simple colonial organisms from truly multicellular ones. Unlike *Gonium, Volvox* cannot be considered simply a colony or cluster of individual cells. It is a single organism whose cells have lost their ability to live independently. If a sufficient number of them become damaged for some reason, the entire sphere of cells will die.

What has *Volvox* gained by this arrangement? In giving up their independence, the cells of *Volvox* have become specialists. No longer does every cell carry out all of life's functions (as in colonial forms); instead, certain cells specialize to carry out certain functions while leaving other functions to other specialists. In *Volvox* this process goes no further than having certain cells specialized for reproduction while others, unable to reproduce themselves, fulfill the needs for photosynthesis and locomotion. In more complex multicellular organisms, such as man, the degree of specialization is carried to greater lengths. Each cell has one or two precise functions to carry out. It depends on other cells to carry out all the other functions necessary to maintain the life of the organism and thus its own. This process of specialization and division of labor among cells is called **differentiation.** One of the great problems of biology is how differentiation arises among cells, all of which, having arisen by mitosis, share the same nuclear controls. We will examine some tentative answers to this problem in Chapter 8.

Specialization and division of labor results in increased efficiency in the same way that a modern shoe factory is more efficient than the individual boot-maker of earlier times. A price must be paid, however, for this increased effi-ciency. If accurate communication and coordination between the various parts of the shoe factory break down, the output of the factory will be adversely af-fected. In multicellular organisms, too, there must be proper communication and coordination between all parts. We do not know just how this is accom-plished in *Volvox*. In the more elaborate multicellular forms, hormones and nerves perform this function.

Our shoe factory analogy can be pushed one step further. If the workers in any one production department go out on strike, the output of shoes from the factory will soon cease. In an analogous way, if one group of specialized cells in a multicellular organism ceases to carry out its function, all the other cells of the organism will be affected. If the human heart fails, every other cell in the body will soon die no matter how vigorously and efficiently each had been func-tioning at the time.

You might well argue that a multicellular organism like man is really no more efficient at carrying out life's activities than a *Chlamydomonas* cell. In the sense that *Chlamydomonas* seems to be in no more (perhaps even less) danger of extinction than man, this is true. However, a comparison of its habitat with that of man may reveal what is really gained by specialization and division of labor. *Chlamydomonas* is almost entirely at the mercy of its environment. About its only means for coping with any adverse change in its environment is to swim away from it or form a resting stage until such time as conditions improve. Many changes in its environmental conditions (e.g., temperature or chemical changes in the water) can cause its death. Thus the habitat of *Chlamydomonas* is a very narrow one and it is at the mercy of it. Man, on the other hand, can live successfully in a great variety of habitats. When conditions change, he is usually able to cope successfully with them. Changes in temperature, diet, etc., have remarkably little effect upon the individual cells of which his body is composed.

We cannot be certain that *Gonium, Pandorina, Eudorina,* and *Pleodorina* represent stages in the evolution of multicellular *Volvox* from unicellular *Chlamydomonas*. Whether they do or not, these organisms illustrate one way in which colonial forms may have arisen from unicellular ones and multicellular forms from colonial ones. They also illustrate the subtle shift in cell relation-ships that occurs as one crosses the uncertain boundary between colonial forms and multicellular, differentiated ones.

One of the principal features of living things is the complex organization of the matter out of which they are made. In these chapters we have examined various levels of this organization from the protons, neutrons, and electrons out of which all matter is made to the structures out of which living cells are made.

The organization of life is not a static, unchanging thing. To be alive is to constantly build, tear down, and rebuild the structural materials out of which the organism is composed. This building up and tearing down depends upon the exchange of materials between the organism and its environment. The exchange of materials and the chemical transformations carried out on these materials within the organism constitute its **metabolism.** It is the topic we will study in Part II.

EXERCISES AND PROBLEMS

1   In what ways do typical plant cells differ from animal cells?

2   List all those structures found in a rat liver cell that seem to be constructed from "unit membranes."

3   What are the advantages of specialization and division of labor among the cells of an organism? What are the disadvantages?

4   Assuming them to be spherical in shape, compare the surface areas of a mycoplasma, staphylococcus and human cheek cell (see Fig. 3–2).

5   How would their *volumes* compare if they were all spherical?

6   What organic molecules are used in the construction of the (a) plant cell wall, (b) chromosome, (c) ribosome, (d) cell membrane, (e) lysosome, (f) oil droplet?

REFERENCES

1   Brachet, J., "The Living Cell," *Scientific American,* Reprint No. 90, September, 1961. Presents an excellent summary of cell structure as revealed by the electron microscope.

2   Swanson, C. P., *The Cell,* 2nd ed., Foundations Of Modern Biology Series, Prentice-Hall, Inc., 1964. The first 4 chapters discuss general cytological techniques and what they have revealed about cell structure.

3   Robertson, J. D., "The Membrane of the Living Cell," *Scientific American,* Reprint No. 151, April, 1962. Discusses the structure of the cell membrane and speculates about how other "unit membrane" structures may have developed from it.

4   Dippell, Ruth V., "Ultrastructure and Function," *This Is Life,* ed. by W. H. Johnson and W. C. Steere, Holt, Rinehart and Winston, New York, 1962. Emphasis on the many roles played in the cell by "unit membrane" systems.

5   Preston, R. D., "Cellulose," *Scientific American,* September, 1957. Helps to bridge the gap between our knowledge of the structure of individual mole-

cules and the cell structures made from them. This article not only describes the structure of individual cellulose molecules but also shows how they are organized to form cell walls.

6   de Duve, C., "The Lysosome," *Scientific American,* Reprint No. 156, May, 1963.

7   Jensen, W. A., *The Plant Cell,* Fundamentals of Botany Series, Wadsworth Publishing Company, Inc., Belmont, California, 1964.

8   Hokin, L. E., and Mabel R. Hokin, "The Chemistry of Cell Membranes," *Scientific American,* Reprint No. 1022, October, 1965. Emphasizes the role of phospholipids in the structure and functioning of the cell membrane.

PART II

ENERGY FLOW IN THE CELL

Plasmolyzed cells in the fresh-water plant *Anacharis densa*, which has been placed in sea water. The space between the cell membranes and the cell walls has become filled with sea water. ▶

CHAPTER 4

THE METABOLISM
OF CELLS

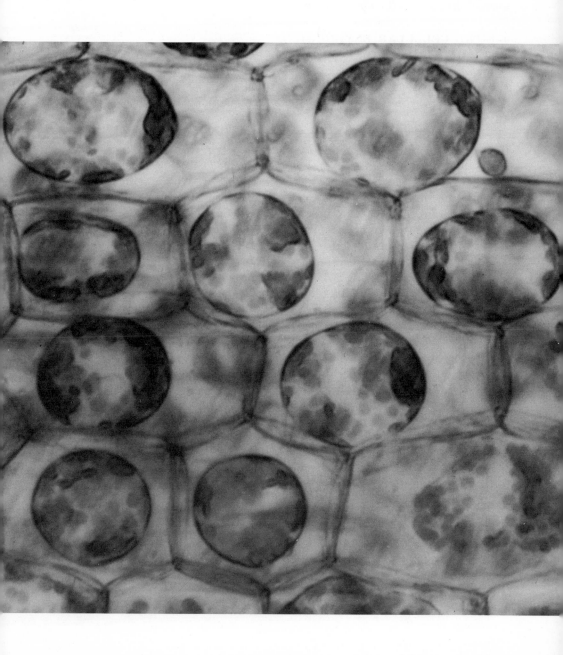

One of the fundamental characteristics of living things is metabolism. Metabolism is the *exchange* of matter and energy between the organism and its environment and the *transformation* of matter and energy within the organism. Each cell, to survive, must secure matter and energy from its environment, transform this matter and energy, and release the waste products of these transformations back to the environment.

4-1 THE CELLULAR ENVIRONMENT

What is the environment of cells? All living cells are bathed by liquid. This statement may seem perfectly obvious when one considers organisms that live in fresh or salt water. But what of the many organisms that live in a terrestrial environment—in the earth, on the surface of the earth, or occasionally flying above the earth? A close examination reveals that in all these cases, too, each living cell is continually bathed in liquid. The skin which man and his fellow vertebrates expose to the air consists of layers of dead cells which protect the living cells beneath from the drying effects of the air. Where living cells are exposed to the environment, as in the epithelium that lines our air passages and the transparent cornea at the front of our eyes, secretory cells bathe the exposed surfaces in a continuous supply of moisture. In a similar way, the external skeleton of insects, the bark of trees, and the waxy covering of leaves all consist of dead cells or waterproof secretions of cells which permit the underlying living cells to remain protected by at least a film of moisture.

Within the soil itself, living cells may be directly exposed to the environment, but here again the environment is liquid. Each soil particle is surrounded by a film of moisture. The delicate root hairs of plants, and the myriads of microorganisms and tiny animals that live in the soil are in contact with this moisture.

In considering the cellular environment of more complex plants and animals, one must consider also the environment of the cells that are not close to the exterior of the organism. These deep-lying inner cells are also in contact with liquid. Sap in plants, the blood of insects, and lymph in man are examples of fluids that bathe the inner cells of higher organisms. Lymph and the blood plasma from which it is derived make up about 20% of the body weight of man. Because these fluids are outside the cells, we will refer to them as the *extracellular fluid* or **ECF.** For the living cells of our body, the ECF *is* the environment. To distinguish between the external environment of our bodies (air) and the actual environment of our cells, the French physiologist Claude Bernard a century ago referred to the latter as the *internal environment.* He studied its properties carefully and found them to be remarkably stable. This is particularly true of the most complex of the multicellular organisms, the birds and mammals. Whether placed in warm locations or cold, whether recently fed or starved, no matter what kind of food taken in the diet, he found that the composition of the ECF remains relatively unchanged. In a real sense, the most adaptable of our higher animals actually are made of cells whose environment, the ECF, remains unchanged despite wide fluctuations in the external environment. Bernard was able to discover several mechanisms by which the mammalian body is able to maintain this constancy of its internal environment. He was so struck by these findings that he wrote: "The constancy of the internal environment is the condition of a free and independent life." Today we use the term **homeostasis** to describe this constancy of the ECF.

In the years since Bernard's discoveries, physiologists have studied the internal environment of other kinds of animals and found that they, too, have mechanisms for maintaining a stable ECF. The less complex animals are, however, less capable in this respect, and this may well account for the more restricted lives that they live.

4-2 THE COMPOSITION OF THE ECF

1. **Water.** The most obvious component of the ECF of any organism—animal, plant, or microorganism—is water. This substance is uniquely suited for the functions it must carry out as the chief component of the ECF. It is unsurpassed as a solvent. In this role it brings to the cell many of the other molecules and ions that make up the ECF and without which life would be impossible. Not only does water serve as the medium in which substances are carried to and away from cells, but it also enters cells and plays a vital role in the chemical activities within.

2. **Gases.** The ECF also contains gases, the most important of which are oxygen and carbon dioxide. Almost all living cells require oxygen and must get rid of their waste carbon dioxide. When chlorophyll-containing cells are exposed to the light, they take carbon dioxide from and release oxygen to the ECF.

3. **Minerals.** The ECF also contains dissolved minerals, or salts, in the form of positively and negatively charged ions. Figure 4-1 lists most of the ions that are found in the ECF of animals and plants. Some, such as calcium (Ca^{++}), potas-

Substantial quantities	"Trace" quantities	
Na^+ (except for plants)	Fe^{++}	
K^+	Cu^{++}	
Ca^{++}	Mn^{++}	
Mg^{++}	Zn^{++}	
$PO_4^{\equiv}$	B^{+++}	— required by plants; certain protists
Cl^-	Mo^+	— required by plants; certain protists and animals
$SO_4^=$	V^{++}	— certain protists and animals
HCO_3^-	Co^{++}	— certain animals, protists, and plants
	I^- $Se^=$ $\}$	— certain animals only

Fig. 4-1

Inorganic ions necessary for most organisms.

sium (K$^+$), sodium (Na$^+$), chloride (Cl$^-$), phosphate (PO$_4^\equiv$), and bicarbonate (HCO$_3^-$), are present in relatively large quantities in animals. Calcium ions are an essential constitutent of our bones as well as of the "cement" that holds individual cells together in tissues. A proper balance of potassium and sodium ions within and without the cell is necessary for the responsiveness shown by nerve and muscle cells. Chloride ions play a major role in maintaining the normal physical properties of the ECF. Phosphate ions, as we shall see, are intimately involved in the distribution and use of energy within the cell. Bicarbonate ions function in the transport of carbon dioxide and help keep the ECF from becoming too acid or too alkaline. More will be said about this function later. Plants require substantial quantities of nitrates (NO$_3^-$) in order to synthesize amino acids and, from them, proteins.

Many other ions are required only in minute quantities. Most of these "trace elements" (e.g., Cu^{++}, Zn^{++}, Mn^{++}, Co^{++}) are either incorporated in enzymes directly or are necessary for the activation of enzymes. Iodine is incorporated in the hormone thyroxin, and small quantities of the fluoride ion (F$^-$) are important in strengthening the mineral portions of teeth and bone.

Not only must certain ions be present in the ECF, but the relative concentration of the various ions must often be held within rather narrow limits. When physiological experimentation is performed with organs or tissues removed from animal bodies, it is necessary to keep the specimen moistened with a solution which contains the same ratio of the major ions (Ca^{++}, Na$^+$, K$^+$) as is found in the ECF of the intact organism. One such solution is called Ringer's solution after the British physiologist, Sidney Ringer, who developed it.

The original discovery, as is so often the case in scientific work, was made as a result of chance. It was the practice at that time to use a simple salt solution (Na$^+$ and Cl$^-$ ions) for bathing tissues. On one occasion, however, a particular batch of salt solution was found to keep frog hearts alive for many times the normal length of time. When it was found that an assistant in Ringer's laboratory had made that particular solution with tap water rather than with distilled water, the search was on for the other necessary ions. Ringer's solution resulted.

4. **Foods and Vitamins.** The ECF also contains organic compounds which serve as food or vitamins for the cells. Food substances serve as a source of energy and as a source of material for growth and repair. They include lipids, nitrogen-containing compounds such as amino acids, and perhaps most widespread of all, a carbohydrate source such as glucose. The blood plasma of man contains 0.1% of glucose or, as it is sometimes called, "blood sugar."

5. **Hormones.** Hormones are an important component of the ECF of vertebrates, higher plants, and perhaps all multicellular organisms. They are chemical substances, released into the ECF by certain cells, which affect the metabolic activities of other cells in the organism.

6. **Wastes.** The waste products of cell metabolism are also components of the ECF. Among the most important to animals are the waste products of protein and nucleoprotein metabolism. These nitrogen-containing wastes, such as ammonia and urea, are somewhat poisonous. Their level in the ECF must not be allowed to exceed certain limits or death will result. In most organisms,

precise homeostatic devices (e.g. the kidney) have evolved which regulate the level of these wastes in the ECF. Although there are some cases where too low a level of a given waste in the ECF would be harmful, the major emphasis is on preventing too great an accumulation of the waste.

7. **pH.** An important requirement for the ECF is that it be neither too acid nor too alkaline. Degree of acidity is measured on a scale of pH units. Pure water, which is neutral, that is, neither acid nor alkaline, has a pH of 7. Acid substances are substances which produce more H⁺ ions than are found in pure water. They have pH values of less than 7. Alkaline (or basic) substances are substances which combine with H⁺ ions, thus leaving fewer than are found in pure water. They have pH values greater than 7 (Fig. 4–2). As we saw in Chapter 1 (see Section 1–9), both OH^- ions and HCO_3^- ions combine with H^+ ions:

$$OH^- + H^+ \rightleftharpoons H_2O$$

$$HCO_3^- + H^+ \rightleftharpoons H_2CO_3 \rightleftharpoons H_2O + CO_2.$$

Thus, when sodium hydroxide (NaOH) or sodium bicarbonate ($NaHCO_3$) is dissolved in water, a basic solution results.

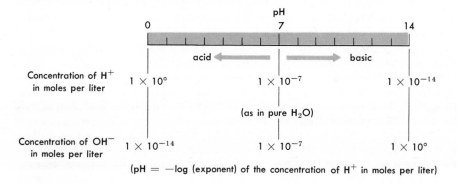

Fig. 4–2
The pH scale. A shift of one pH unit represents a tenfold shift in acidity. A solution of pH 7 is neutral.

The pH scale is a logarithmic one: each unit on the pH scale represents a tenfold increase or decrease in the concentration of H^+ ions over the next lower or higher unit. For example, a solution with a pH of 5 is ten times more acid (has a concentration of H^+ ten times greater) than a solution of pH 6.

Most living cells are extremely sensitive to changes in the pH of their ECF. The pH of human plasma is usually maintained at a value between 7.34 and 7.44. If these limits should be greatly exceeded (below pH 6.8 or above pH 7.8), serious illness and perhaps death would follow. Fortunately, our plasma is well supplied with substances called **buffers,** that act to prevent any sudden shift in pH. The proteins found in blood plasma act as buffers. If the pH should start to decrease (violent exercise or holding the breath will cause this), the proteins combine with the new H^+ ions, thus keeping the pH constant. If the pH should start to rise, the proteins release H^+ ions to the ECF, thus restoring the original condition. The plasma proteins thus serve as a homeostatic mechanism for maintaining a constant pH in the ECF.

Blood plasma also contains bicarbonate ions (HCO_3^-) which serve in the same capacity. In the presence of an increased concentration of H^+ ions (and thus a decrease in pH), the following reaction occurs:

$$HCO_3^- + H^+ \rightleftharpoons H_2CO_3.$$

When the H_2CO_3 reaches the lungs, it decomposes into carbon dioxide (which passes into the air) and neutral water:

$$H_2CO_3 \rightleftharpoons H_2O + CO_2.$$

Thus the danger of too many H^+ ions in the ECF is neatly avoided.

8. **Temperature.** Another important characteristic of the ECF is its temperature. Microorganisms have no control over the temperature of their ECF. It is simply a function of the climatic conditions around them. This is generally true of plants, invertebrate animals, and the so-called cold-blooded vertebrates: the fishes, amphibians, and reptiles. In some of these cases, however, the organisms have a certain degree of control over the temperature of their ECF. When the temperature in a beehive becomes too high, the workers begin cooling the hive by bringing in water and fanning the air with their wings. Fanning brings in fresh air and also speeds the evaporation of the water, which is itself a cooling process.

Goldfish have been shown to seek water in a preferred temperature range. In fact, they have been trained in the laboratory to trigger the addition of cold water when their surroundings get too warm.

Many lizards also exert considerable control over the temperature of their ECF at least for the periods when they are active. Basking in the sun enables a lizard to raise its body temperature to the preferred range for its species even though the temperature of the surrounding air may be considerably cooler. When the air temperature becomes quite high, lizards can still maintain their preferred temperature by evaporating water from their tongue (panting) and retreating to shady locations.

It is among the birds and mammals, however, that we find the strictest control over the temperature of the ECF. Birds and mammals are able to maintain the temperature of their ECF within very narrow limits despite wide fluctuations in the temperature of the surrounding air. For this reason, they are often called "warmblooded," or better, **homeothermic.** A healthy human is capable of maintaining the temperature of his ECF within a degree or so of 37.5° C (98.6° F) at rest or during violent exercise, in warm surroundings or cold.

EXCHANGE OF MATERIALS

At the beginning of this chapter it was stated that metabolism involves the exchange of matter and energy with the environment and the transformation of this matter and energy within the cell. The environment of an individual cell is its ECF. We must now examine the process by which substances are exchanged between it and the cell.

4-3 DIFFUSION

Some of the materials dissolved in the ECF enter the cell by diffusion across the cell membrane from the ECF into the cytoplasm. Similarly, other substances diffuse out of the cell into the ECF.

Fig. 4-3

Diffusion. Even in the absence of air currents, the perfume molecules will eventually become evenly distributed throughout the room. Their own random motion accounts for this.

    To help us understand the process of diffusion, consider what happens when a bottle of perfume is opened in one corner of a sealed room. Even with air currents completely eliminated, the odor of the perfume gradually spreads to every part of the room. Eventually the bottle of perfume becomes empty and the intensity of the odor is uniform throughout the room.

    What has happened? The liquid perfume has evaporated. The molecules of which the perfume is composed have not, however, disappeared. At first they mix with the molecules of air in the vicinity of the bottle. Then, as time goes on, they spread to all parts of the room (Fig. 4-3). This process of diffusion continues until the molecules are uniformly distributed throughout the room. We can define diffusion as the net movement of molecules or ions from a region where they are in high concentration to a region of lower concentration.

What force accounts for this motion? It is believed that all the molecules in a gas (such as air) or in a solution are in constant random motion. This motion takes place in straight lines and continues in a given direction until one molecule collides with another and rebounds in a different direction.

One visible bit of evidence to support this theory is the phenomenon known as Brownian motion. If with a high-power microscope you examine a water suspension of tiny particles such as bacteria, you will see that the particles are in continual random motion or vibration. This motion is named after the English botanist Robert Brown, who first described it in 1827. It should not be confused with any swimming motions displayed by some bacteria. It is believed to be caused by continual collisions between the water molecules and the bacteria. It is thus an entirely physical process. Dead bacteria show Brownian motion just as well as living ones do. Why, though, do you suppose larger objects fail to show this activity?

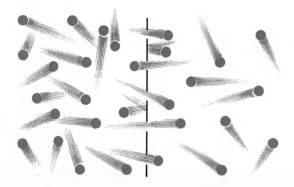

Fig. 4-4

Diffusion through a barrier. As long as there is a greater concentration of molecules on the left, more molecules will pass from left to right than from right to left.

The odds are that any object moving about entirely at random will gradually move away from its starting point. Furthermore, in the case of the perfume bottle, any molecule moving away from the bottle would be apt to travel farther than a molecule moving back towards the bottle. This is because there are more perfume molecules nearer the bottle and hence a greater chance of collisions to occur. Thus there will be a net movement of the molecules from the region of greater concentration to the region of lesser concentration. When the molecules are finally distributed evenly throughout the room, the process of diffusion ceases. This does not mean that motion of the molecules ceases. There is no longer a net movement or trend, however, and hence there is no more diffusion.

In the case of the perfume bottle in the room, no barrier was placed between the region of high concentration and the region of low concentration. In the case of diffusion between cells and their ECF, a definite barrier, the cell membrane, is present. However, as long as there are pores in the membrane sufficiently large to let a given size of molecule pass through, diffusion of that substance can occur. A glance at Fig. 4–4 should make the mechanism clear. Whenever a small molecule strikes the macromolecules of which the membrane is made, it rebounds. If, however, it reaches the membrane in the vicinity of a pore, it can pass through to the other side. With a greater concentration of a given type of molecule on one side of the membrane than on the other, there will be more collisions with the membrane on that side. There will also be more successful passages through the membrane in this direction. Although molecules do pass through the membrane in both directions, diffusion is considered to occur only in the direction of greater movement.

4–4 OSMOSIS

Osmosis is simply a special case of diffusion. Chemists define osmosis as the diffusion of any solvent through a differentially permeable membrane. Cell membranes, having pores which permit the passage of some molecules but prevent the passage of others, are differentially permeable. The universal solvent in living things, as was mentioned earlier, is water. For our purposes, therefore, osmosis can be defined as the diffusion of water through a differentially permeable membrane from a region of high concentration to a region of low concentration. Note that concentration refers to the concentration of the solvent, water, and not to the concentration of molecules or ions which may be dissolved in the water. The exchange of water between the cell and its environment is such an important factor in cell function that it justifies the special name of osmosis.

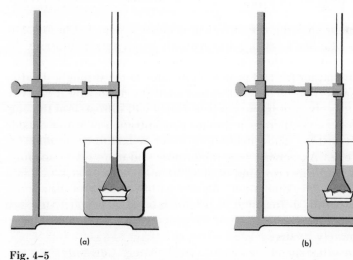

(a)   (b)

Fig. 4–5

Osmometer. (a) At start. (b) A few hours later. The cellophane is a differentially permeable membrane; water molecules pass through it more readily than do sugar molecules.

An experiment that demonstrates osmosis is shown in Fig. 4–5. The lower opening of the glass tube is covered with a sheet of cellophane. This acts as a differentially permeable membrane, permitting the rapid passage of water molecules but obstructing the passage of larger molecules. The interior of the tube is filled with molasses, a concentrated solution of sugar in water. The whole apparatus is placed in a beaker containing distilled water. In which direction will osmosis occur? The water concentration in the beaker is 100%. The water concentration inside the tube is less than this because a given volume of molasses contains fewer water molecules than the same volume of distilled water. There is, therefore, a net movement of water molecules through the cellophane membrane and into the tube.

As additional water molecules enter the tube, the volume of fluid increases. The molasses is forced up the tube. This force arises from pressure exerted by the diffusion of water molecules into the tube. The pressure is called osmotic pressure. The greater the difference in water concentration on either side of the membrane, the greater the tendency for osmosis to occur and thus the greater the osmotic pressure. In fact, when the column of molasses stops rising, we have a rough measure of the osmotic pressure of the system. The weight of the column of water finally counterbalances the osmotic pressure and osmosis ceases. Note that the water concentrations on the two sides of the membrane are still not equal by any means. However, the increase of pressure on the inner surface of the membrane, created by the weight of the column of molasses above, causes water molecules to be forced or filtered back out through the pores. When the rate at which this filtration process occurs becomes equal to the rate at which water molecules are coming in because of the difference in concentration, osmosis ceases.

Another experiment showing the effects of osmosis is depicted in Fig. 4–6. The shell membrane of a hen's egg is a differentially permeable membrane. Its pores are large enough to allow the easy passage of water molecules, but are not large enough to allow larger fat and protein molecules to get through. If after carefully removing a portion of the waterproof shell (by dissolving it away in dilute acid), one places the egg in pure water, water will diffuse into the egg. This osmosis occurs because there is a greater concentration of water outside the egg (100%) than within. The surrounding water is said to be **hypotonic** to the contents of the egg. As osmosis continues, more and more water accumulates within the egg and this crowding in of additional molecules results in a buildup of pressure. The development of pressure within a cell (and the egg *is* a single cell) as a result of the diffusion of water into it is called **turgor.** Although the remaining portion of the shell can resist the pressure, the unprotected membrane cannot. Ultimately it bursts.

What of organisms that spend their lives in fresh water? Certainly the water concentration of their cytoplasm can never approach that of the pure water surrounding them. In the case of the cells of freshwater plants, water passes into the cell by osmosis and turgor quickly develops. The strong cellulose walls

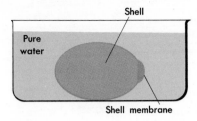

Fig. 4–6

Egg osmometer. The pores of the shell membrane permit water molecules to diffuse into the egg but are too small to let the macromolecules within the egg diffuse out. Consequently, the volume of materials within the egg increases until increasing pressure ruptures the membrane.

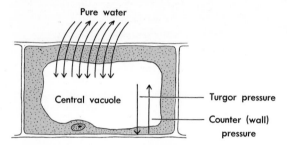

Fig. 4–7

When the pressure within the cell finally equals the osmotic pressure, the movement of water in and out of the cell will reach equilibrium.

of the cells are, however, capable of withstanding this pressure. Soon the pressure within the cell equals the osmotic pressure even though the two water concentrations are not equal. At this point, osmosis ceases (Fig. 4–7).

Fresh-water animals and protozoans lack cellulose cell walls so they must cope differently with life in a hypotonic environment. Water enters their cells continually by osmosis but only slight turgor can safely be developed. The problem is solved by employing energy and some contractile structure to pump the excess water back out into the environment. The single-celled amoeba accomplishes this by means of a contractile vacuole (Fig. 4–8), in which water entering the cell by osmosis is collected. When the vacuole is filled, the amoeba contracts it forcing the water out through a pore which forms momentarily in the cell membrane. Note that the water, which had entered the cell as a result of the random molecular activity of osmosis, leaves the cell by flowing out because of a force generated by the cell. The creation of this force requires the expenditure of energy by the cell.

Life in the oceans involves quite different osmotic conditions than life in fresh water. Sea water contains about 3.5% of various ions, especially Na^+ and Cl^-. This results in a water concentration which is approximately the same as

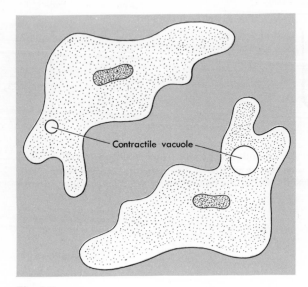

Fig. 4–8

The amoeba bails out the continual influx of water from its hypotonic surroundings by alternately filling and emptying its contractile vacuole. The contents of the vacuole may be discharged at any point on the surface of the cell.

that in the cytoplasm of marine plants and the invertebrate animals that live in the sea. Consequently, these organisms are able to exist in a state of equilibrium with respect to the water in their surroundings. They neither gain nor lose water by osmosis. We say that sea water is **isotonic** to their cytoplasm.

That a given volume of sea water is about 3% salt by weight, and thus 97% water, does not mean that the water concentration of the cytoplasm of these marine plants and animals is also 97% by weight. The speed with which diffusion or osmosis takes place is a measure of the difference in the number, not weight, of the molecules or ions involved. When osmosis occurs, the region of higher water concentration is simply the region which contains the greater number of water molecules in a given volume of the mixture. The cytoplasm of these marine organisms may contain as little as 80–90% water by weight. However, much of the remaining material in the cytoplasm consists of protein. These macromolecules make up a substantial fraction of the weight of the cytoplasm but contribute only a minor osmotic effect because of the relatively small number of molecules involved. Similarly, a 0.9% salt solution (99.1% water) is isotonic to human blood plasma although the latter contains only 90% water by weight. The number of water molecules in a given volume of each of the two solutions is, however, the same.

When a fresh-water (or terrestrial) plant is placed in sea water, its cells quickly lose turgor and the plant wilts. This is because a given volume of seawater contains a smaller number of water molecules than a given volume of the cytoplasm of these plants. The sea water is said to be **hypertonic** to the cytoplasm. As water continues to diffuse from the cytoplasm into the sea water, the protoplast gradually shrinks. This condition is known as **plasmolysis.** Note in the chapter opening illustration how the protoplasts have pulled away from the cell walls, which still retain their original shape.

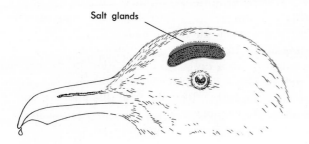

Fig. 4–9

The salt glands of the herring gull. The fluid excreted by the glands is saltier than the blood.

The ECF of bony fishes has a water concentration which is substantially higher than that of sea water. Thus they live in a medium which is hypertonic. Whereas fresh-water organisms have to cope with water passing continually into their body by osmosis, the salt-water bony fish continually loses water by osmosis. Once again, however, survival depends upon the expenditure of energy to combat the force of osmosis. The fish drinks sea water and then uses metabolic energy to desalt it. The salt is transported back into the environment at the gills. Marine birds, which sometimes pass long periods of time away from fresh water, and sea turtles use a similar device. They, too, drink salt water to take care of their water needs and use metabolic energy to desalt it. The salt is extracted by two glands in the head and released (in a very concentrated solution) to the outside through the nostrils (see Fig. 4–9).

4–5 ACTIVE TRANSPORT

The ability of marine bony fishes and marine birds to transport ions from a region where they are in low concentration (the ECF) to a region where they are in high concentration (the ocean and the ducts of the salt glands, respectively) implies that some force other than diffusion is at work. In both cases, the transport of ions occurs against the concentration gradient and hence in a direction opposite to that which would occur by diffusion alone. This movement of ions or molecules against a concentration gradient is known as active transport.

It is described as active because the cell must expend energy to accomplish the transport against the passive force of diffusion. The ability of cells to accomplish active transport of ions and molecules to or from the ECF is a widespread one. Marine organisms frequently have certain ions in their cytoplasm in concentrations a thousand or more times greater than in the surrounding sea water. Iodine in brown algae and vanadium in certain invertebrates are found concentrated in this way. This must mean that these cells have managed to transport ions actively from the sea water to their cytoplasm.

The cells lining the intestine of the rat can transport glucose actively from a lower concentration in the intestine to a higher concentration in the blood. Most cells actively transport sodium ions out of the cell and potassium ions into the cell. The transmission of nerve impulses depends upon this phenomenon. The filling of the amoeba's contractile vacuole requires the active transport of water molecules from the cytoplasm into the vacuole itself.

The mechanism by which active transport is accomplished is not yet clearly understood. There is no doubt that the cell must use some of the energy produced during its metabolism to carry out active transport. Anything that interferes with the cell's production of energy also interferes with active transport. It is probable that specific enzymes are required which serve to carry a given molecule or ion from one side of the cell membrane to the other. In any case, the plasma membrane cannot be considered as simply a passive barrier to the diffusion of molecules. It may exert a very decided influence on what substances pass through it.

4-6 PHAGOCYTOSIS AND PINOCYTOSIS

Still another mechanism by which the cell transports materials from the ECF into the interior is to engulf them by folding inward a portion of the cell membrane. The pouch that results then breaks loose from the outer portion of the membrane and forms a vacuole within the cytoplasm. In some cells, solid particles may be engulfed in this manner. The process is called **phagocytosis** or "cell eating." The amoeba derives its nourishment by ingesting smaller microorganisms in this fashion (Fig. 4-10). Phagocytic cells in the blood of many animals serve the extremely valuable function of engulfing foreign particles, such as bacteria, that may get into the animal's body. Phagocytosis also permits any molecules present in the surrounding medium to which the cell membrane is normally impermeable to gain entry into the cell along with the ingested food particles.

The ability to engulf solid materials is found in only a few kinds of cells. Many cells, however, are able to carry on a similar action called **pinocytosis.** In pinocytosis ("cell drinking"), the cell engulfs droplets of the surrounding ECF by a mechanism quite similar to that of phagocytosis, although the pockets formed by the cell membrane are smaller. Figure 4-11 is an electron micro-

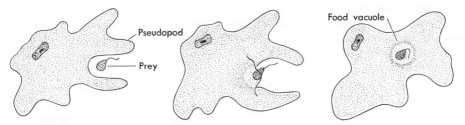

Fig. 4–10

Phagocytosis in the amoeba. Even when totally engulfed, the victim is separated from the cytoplasm of the amoeba by a membrane around the food vacuole.

graph of a section of the wall of a capillary (the smallest type of blood vessel in our bodies). At the top is the interior or bore of the capillary. In the middle is the tissue space separating the capillary wall from a nearby muscle cell (bottom). The small inpocketings of the cell membrane are clearly seen. Most of these are engulfing the ECF of the tissue space, but some can also be seen on the other side of the wall, apparently engulfing fluid from within the capillary.

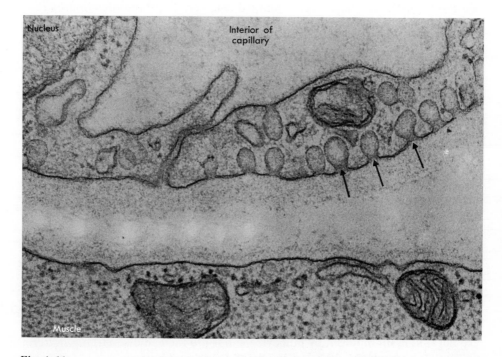

Fig. 4–11

Pinocytosis in the thin cell that forms the wall of a capillary. Note that the inpocketings have formed on both surfaces of the cell. (67,500 X, from Fawcett: *The Cell: Its Organelles and Inclusions,* W. B. Saunders Co., 1966.)

Pinocytosis requires the expenditure of energy by the cell and enables materials to enter the cell even though they are present in the ECF in a concentration less than that within the cell. In this sense, then, pinocytosis is a special case of active transport, although it is probably not the only mechanism by which active transport occurs.

While the size of the pockets formed in phagocytosis and pinocytosis is quite different, there is evidence to suggest that fundamentally the processes are similar. In cells such as the amoeba, where both phagocytosis and pinocytosis can occur, either activity temporarily inhibits the other. Presumably, an amoeba has only a certain amount of cell membrane to devote to this inpocketing activity at any one time. An amoeba which has been engulfing food particles by phagocytosis is temporarily unable to carry on as much pinocytosis as usual. The reverse is also true.

The mere act of folding in a bit of the cell membrane with its content of materials from the ECF does not actually get these materials into the cytoplasm. They are still retained within vacuoles which, although surrounded by cytoplasm, are separated from it by a membrane. In the amoeba, the vacuole formed as a result of phagocytosis remains intact until it is finally discharged at the margin of the cell. During the period between its formation and discharge, however, some substances are certainly exchanged between it and the surrounding cytoplasm. Digestive enzymes enter the vacuole. This probably occurs when lysosomes (see Section 3–11) fuse with it. The small molecules produced by digestion pass through the vacuolar membrane into the cytoplasm. As for pinocytotic vacuoles, a similar mechanism may be at work. There is good evidence that digestion of macromolecules occurs within them. The products of digestion (e.g. glucose) quickly enter the cytoplasm from the vacuoles although these same substances cross *cell* membranes far more slowly.

4–7 REVERSE PHAGOCYTOSIS AND PINOCYTOSIS

Just as materials can also *leave* cells by diffusion and active transport, so they may also leave by reverse phagocytosis or pinocytosis. The discharge of the old food vacuole of an amoeba, mentioned above, is an example of reverse phagocytosis. In cells that secrete large amounts of protein, the protein first accumulates in a membrane-bounded sac within the Golgi complex. This moves to the surface of the cell where its membrane fuses with the cell membrane, and it discharges its contents to the outside (Fig. 4–12).

Reverse pinocytosis has also been observed. The cells lining our intestine synthesize tiny droplets of fat and then discharge these from the cell by reverse pinocytosis. It may even be that some of the tiny vacuoles shown in Fig. 4–11 are not taking up material by pinocytosis, but are instead discharging material by reverse pinocytosis. In other words, the pinocytotic vacuoles formed at one surface of the cell may, after being detached, move through the cell to the other surface and there discharge their contents. In this way, materials can be moved efficiently through the capillary wall.

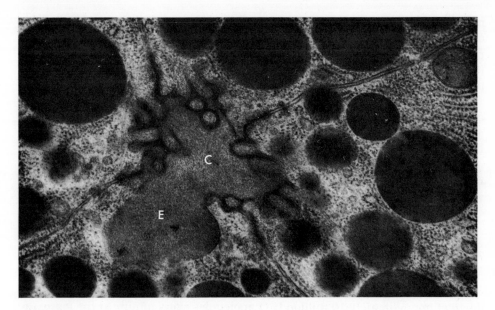

Fig. 4-12
Reverse phagocytosis. The large spherical bodies contain digestive enzymes. One is seen in the process of discharging its contents (E) into the intercellular canal (C) where the corners of these four bat pancreas cells meet. (30,000 X, from Fawcett: *The Cell: Its Organelles and Inclusions,* W. B. Saunders Co., 1966.)

CELL CHEMISTRY

The exchange of materials between a cell and the ECF is one aspect of metabolism. The chemical transformation of these materials within the cell is another. Literally hundreds of the chemical changes that occur within cells have been discovered and the list grows longer every year. In the following chapters several of the most common and most important of these will be described. Despite their variety, we will find that there is one feature they all share: each chemical change can occur only in the presence of a specific enzyme.

4-8 ENZYMES

In Chapter 1 we saw that even energy-yielding reactions, such as the burning of a strip of magnesium, require the input of some energy before they can proceed spontaneously. This energy is known as the energy of activation. Glucose exposed to oxygen in the air can become oxidized to carbon dioxide and water, but at room temperature the process goes on so slowly as to be imperceptible. When glucose is heated in a flame, however, a greater number of molecules receive their necessary energy of activation and the process proceeds more rapidly. Once a large number of molecules are being oxidized, the energy liberated supplies nearby molecules with their energy of activation. Soon the process becomes self-sustaining and the sugar burns rapidly.

The temperature necessary for this to occur is much higher than that which can be tolerated within living cells. Nevertheless, the oxidation of glucose can go on within living cells at a rapid rate and still at a low temperature. How is this accomplished? In Chapter 1 we saw that substances called catalysts are able to lower the activation energy of molecules and thus permit them to undergo rapid chemical change at lower temperatures. Respiration in living cells likewise requires the activity of catalysts to accomplish this same function. These catalysts are called **enzymes.**

Enzymes, like other types of catalysts, do not participate permanently in the chemical change that they promote. When the chemical change is completed, the enzyme is released unchanged. Because enzymes are not used up by the processes they catalyze, they may be re-used again and again. A single molecule of the enzyme catalase (see Section 1–11) is capable of catalyzing the breakdown of approximately 10 million H_2O_2 molecules every minute at the temperature of ice water (0° C)! Because enzymes can be re-used, only small amounts of each enzyme are needed by a single cell. Since a single cell may be capable of carrying out over a thousand different chemical reactions, and each one of these probably requires its own special enzyme, it is fortunate that large quantities of enzymes are not necessary.

The fact that each chemical reaction in the cell requires its own special enzyme and, conversely, that a single enzyme can catalyze only one (or sometimes a few quite similar) chemical reactions requires some explanation. One theory which has been created to explain this specific relationship between an enzyme and the substance upon which its acts (its **substrate**) is called the **lock-and-key theory.** According to this theory, an enzyme actually combines with the molecule of its substrate for a brief period of time. Presumably the combination requires that each molecule have a shape complementary to that of the other. This would explain the specificity of enzymes. Different substrate molecules each possess a different shape and thus require a different enzyme. The combination between enzyme and substrate makes the substrate molecule more reactive; that is, it lowers the energy necessary for the substrate molecule to undergo chemical change. Once the chemical transformation in the substrate molecule is completed, the products leave the enzyme, which then is free to repeat the process with a new substrate molecule. Figure 4–13 shows how this process is thought to work.

There is quite a bit of experimental evidence to support the idea that the specific action of enzymes depends upon a brief union between the enzyme molecule and the substrate molecule. One type of evidence is the phenomenon of competitive inhibition of enzymes. Certain poisons exert their influence by

Fig. 4-13

The "lock-and-key" theory of enzyme action.

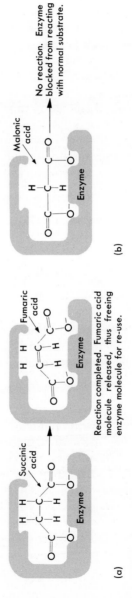

(a)

(b)

Fig. 4-14

(a) Schematic representation of the action of the enzyme, succinic dehydrogenase, on succinic acid. (b) The competitive inhibition of the enzyme by malonic acid.

blocking normal enzyme activity. It is thought that the shape of the poison molecule is so similar to that of the normal substrate molecule that the enzyme combines with it instead. However, the enzyme is not able to promote the chemical change of the new molecule. The poison molecule remains attached to the enzyme molecule, which is therefore no longer able to carry out its normal function. Among our vital enzymes is one that catalyzes the oxidation of succinic acid, a reaction which is an essential part of cellular respiration. The compound malonic acid blocks the activity of this enzyme. The molecular structure of malonic acid is sufficiently similar to that of succinic acid (Fig. 4–14) that it presumably combines with the enzyme but cannot then be transformed and released. The enzyme molecule is thus kept from combining with its normal substrate.

One difficulty that has been encountered in trying to support this picture of enzyme action has been the problem of determining what the actual shape of any enzyme molecule is. All enzymes are proteins and hence exceedingly complex molecules. Only recently has the chemical and physical analysis of certain proteins begun to reveal their shapes. One of these is an enzyme, lysozyme, that is found in egg white and such body secretions as tears. It has a strong antibacterial action because its substrate is the polysaccharide that makes up the cell walls of bacteria.

The lysozyme molecule appears to be roughly globular in shape but with a long, narrow cleft on one surface. The long cellulose-like molecule (Fig. 2–5) of its substrate fits into this cleft. When in the cleft, one of the sugar units in the molecule is twisted out of its normal position. This seems to impart the necessary energy of activation, and the bridge between that sugar residue and the next is broken. In this way, the structure of the bacterial cell wall is broken down. (You can read an account of these discoveries in Reference 9 listed at the end of the chapter.)

So many enzymes have been discovered in living cells that some uniform system of naming them has become essential. Some enzymes are named by using a prefix taken from the name of the substrate and the suffix -ase. Thus fat-digesting enzymes are called lipases, and protein-digesting enzymes are proteinases. Some of the more recently discovered enzymes are named by giving the full name of the substrate, followed by the action carried out on the substrate and then the suffix -ase. Thus the enzyme which oxidizes triose phosphate (by removing two hydrogen atoms) is called triose phosphate dehydrogenase. Some of the enzymes that were discovered early in the history of enzyme chemistry still retain their early names and may not adhere to the principles just described. The protein-digesting enzymes pepsin and trypsin are two examples in this category.

All of the enzymes that have been discovered so far are proteins and are quickly destroyed by high temperatures. Just as boiling water will denature or "cook" the proteins in an egg (see Section 2–3), so it will quickly inactivate enzymes. Enzymes are also sensitive to change in pH. Each enzyme operates most effectively at a certain pH, its activity diminishing at values above or below that point. The enzyme pepsin works most effectively at a pH of 1–2 while the related enzyme trypsin is quite inactive at that pH but functions effectively at a pH of 8.

Many enzymes require for their action some additional substance. This accessory substance may be a relatively small molecule attached firmly to the protein itself. It is then called a prosthetic group. Some of our vitamins (e.g. riboflavin) function as the prosthetic group of certain enzymes. Some trace element ions such as Zn^{++}, Cu^{++}, and Co^{++} are also found in the prosthetic group of enzymes. Sometimes enzyme activity requires only that the accessory substance unite briefly with the enzyme molecule, the rest of the time simply being present in the surrounding medium. Such substances are called coenzymes. Some of our vitamins function as parts of coenzymes. Some trace element ions (e.g., Mg^{++}) need only be present in the surrounding medium for enzyme activity to go on.

Considering the large number of enzymes in the living cell, one might well ask how their activity is regulated. A number of mechanisms play a part in making enzyme action within the cell efficient and well coordinated.

For those enzymes, such as proteinases, which can attack the very substance of the cell itself, we find that their action is inhibited while they are present within the cell. The proteinase pepsin, for example, is manufactured by the cell in an inactive form, pepsinogen. Only when exposed to conditions of low pH outside of the cell is the inhibiting portion of the enzyme molecule removed and the active pepsin produced. Other potentially destructive enzymes are sequestered in the lysosomes (see Section 3–11) and thus isolated from the rest of the cell.

Many of the cell's enzymes cannot move freely within the cell but are, instead, arranged in definite patterns. The enzymes within mitochondria and chloroplasts appear to be organized spatially in such a way that they interact with the greatest efficiency. It is quite likely that spatially organized enzyme molecules are also present on the cell membrane and the membranes of the endoplasmic reticulum.

The activity of enzymes within the cell is also closely regulated by the need for them. If the product of a series of enzymatic reactions (e.g., an amino acid) begins to accumulate within the cell, it specifically inhibits the action of the first enzyme involved in its synthesis. Thus further production of that amino acid is temporarily halted. On the other hand, the accumulation of a substance within the cell may specifically activate the enzyme for which it is the *substrate*. This action also reduces its concentration to normal levels.

The mechanisms mentioned above ensure that the activity of enzymes already present in the cell will be properly regulated. What of enzymes that may not be needed at all or that may be needed but are not present? Here, too, delicate controls are at work. These regulate the rate at which new enzymes are synthesized. If, for example, excess quantities of an amino acid are supplied to a cell from its ECF, the synthesis of all the enzymes by which the cell ordinarily would produce that amino acid for itself will be halted. Thus the cell enhances its efficiency by not producing enzymes that it does not need. Conversely, if a new substrate is made available to the cell, it will stimulate the synthesis of the enzymes needed to cope with it. Yeast cells do not ordinarily ferment the disaccharide lactose and ordinarily no lact*ase* can be detected within their cells. If grown in solutions containing lactose, however, they eventually begin producing lactase and begin to metabolize the sugar.

In these cases where the *synthesis* of enzymes is being regulated, it is clear that this regulation works through the hereditary controls coded in the DNA of the nucleus. The mechanisms by which portions of the hereditary code are thought to be turned on and off in response to the needs of the cell will be examined in Chapter 8.

Whether acting on the enzymes already present within the cell or on the rate of synthesis of new enzymes, it is clear that these control mechanisms work together to stabilize the levels of substrates and products in the cell. In a sense, then, these mechanisms are homeostatic devices, working within the cell, that regulate enzyme activity with the utmost efficiency and in harmony with the changing needs of the cell.

Metabolism in plants and algae includes the manufacture, by photosynthesis, of organic molecules from inorganic ones taken in from the environment. The nutrition of these organisms is said to be **autotrophic.** Animals and all the other organisms that lack chlorophyll must secure their organic molecules from the environment. This form of nutrition is referred to as **heterotrophic.** The way in which heterotrophic organisms extract energy from the organic molecules upon which they feed is the topic of the next chapter.

EXERCISES AND PROBLEMS

1 Describe three different ways in which materials in the human intestine might enter the cells lining it.

2 How can coenzymes be separated from their enzymes?

3 After the molasses reaches its maximum height (see Fig. 4–6), what will happen? Why?

4 Which will have the higher osmotic pressure, a 1-molar solution of glucose or a 1-molar solution of salt? Why?

5 A single molecule of catalase can decompose ten million molecules of H_2O_2 each minute at 0° C. What would you expect the approximate figure to be at 10° C?

6 What would happen if (a) red blood corpuscles, (b) plant cells, (c) an amoeba were placed in distilled water? Explain.

7 What would happen if (a) red blood corpuscles or (b) plant cells were placed in sea water? Explain.

8 What would happen if an amoeba were placed in an isotonic solution?

9 Distinguish between a coenzyme and a prosthetic group.

10 If a mole (28 g) of nitrogen molecules (N_2) and a mole (28 g) of ethylene molecules (CH_2CH_2) were released on opposite sides of a partition that divided a room into equal volumes and then the partition were removed, do you think diffusion of the molecules would occur? Explain.

REFERENCES

1 Wolf, A. V., "Body Water," *Scientific American,* Nov., 1958. Discusses the distribution of water in the body and the physiology of water balance.

2 Anderson, A. J., and E. J. Underwood, "Trace-Element Deserts," *Scientific American,* Jan., 1959. Discusses the importance of trace elements to plants and animals and tells how the addition of small quantities of these substances to the soil is bringing fertility to formerly unproductive regions of Australia.

3 Bogert, C. M., "How Reptiles Regulate Their Body Temperature," *Scientific American,* April, 1959.

4 Holter, H., "How Things Get into Cells," *Scientific American,* Reprint No. 96, September, 1961. An excellent review of the forces of passive and active transport across cell membranes.

5 Solomon, A. K., "Pores in the Cell Membrane," *Scientific American,* Reprint No. 76, December, 1960. A beautiful illustration of the way physical analysis and experimentation can unlock the secrets of biological function.

6 Schmidt-Nielsen, K., "Salt Glands," *Scientific American,* January, 1959. Shows how active transport enables marine reptiles and birds to desalt the sea water they drink.

7 Rustad, R. C., "Pinocytosis," *Scientific American,* April, 1961. Shows how molecules too large to enter the cell by diffusion can be actively engulfed by the cell.

8 Buchner, E., "Alcoholic Fermentation without Yeast Cells," *Great Experiments in Biology,* ed. by M. L. Gabriel and S. Fogel, Prentice-Hall, Inc., 1955. A report of the first demonstration (accidental) that the biochemical activities of a cell may be duplicated by enzymes extracted from the cell.

9 Phillips, D. C., "The Three-dimensional Structure of an Enzyme Molecule," *Scientific American,* Reprint No. 1055, November, 1966. Describes how the shape of the enzyme lysozyme was worked out and how this shape accounts for the antibacterial action of the enzyme.

10 Changeux, J.-P., "The Control of Biochemical Reactions," *Scientific American,* Reprint No. 1008, April, 1965.

CHAPTER 5

ENERGY RELEASE IN THE CELL

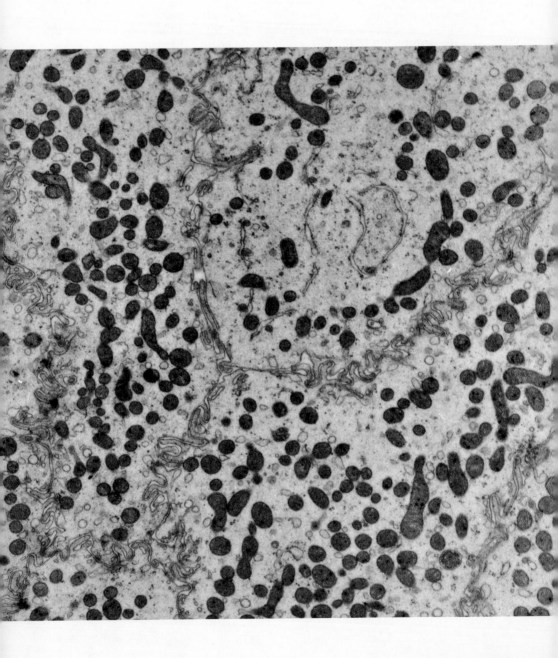

All organisms require a fairly steady supply of materials and energy from the environment in order to stay alive. For many, the chief supply of materials and the only supply of energy come from fairly complex, energy-rich, organic molecules secured directly or indirectly from the environment. (What is the ultimate source of these molecules?) Nutrition that involves dependence upon preformed organic molecules is called **heterotrophic** nutrition, and the organisms using this kind of nutrition are called heterotrophs. The nonchlorophyll-containing microorganisms, the few nongreen plants, and all animals are heterotrophic.

5–1 REQUIREMENTS

The organic molecules that serve as a source of material and energy are the sugars, amino acids, fatty acids and glycerol, and (for purposes of synthesis only) the vitamins. Not all heterotrophs depend on all these organic molecules. Some microorganisms, for example the bacterium *E. coli,* thrive with just sugar as their source of energy. They must, however, take in some inorganic materials such as nitrates in order to synthesize all their other organic constituents. Man is quite demanding in his requirements for preformed organic molecules. He requires carbohydrates, eight of the twenty amino acids found in his body proteins (these are the "essential" amino acids—from them he can synthesize the other twelve), and probably eight or more vitamins. Although he can manufacture fats from sugars, he seems to need certain special fats in his diet also.

5–2 INTRACELLULAR DIGESTION

Solid food materials are usually broken down into a solution of relatively small, soluble, organic molecules before they can be used by heterotrophic organisms. This breakdown process is called **digestion.** In some heterotrophic organisms, digestion is intracellular, that is, it occurs after the solid material has actually been engulfed by a cell.

The *Amoeba* engulfs solid particles such as small protozoans by phagocytosis (see Section 4–6). The prey, with a small amount of the ECF, is incorporated into a **food vacuole** within the cytoplasm of the amoeba. Next, the digestible portions of the food material are digested by enzymes secreted into the vacuole from lysosomes that fuse with it (Fig. 5–1). The soluble food molecules then

◄ Mitochondria in epithelial cells of the gall bladder of a rabbit (15,000 X). In these structures the energy of food is made available to the cell. (Courtesy Dr. Gordon Kaye, Columbia University, College of Physicians and Surgeons.)

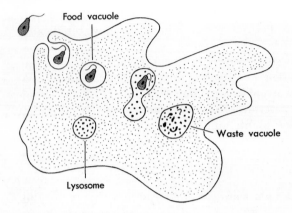

Fig. 5-1

Intracellular digestion in the amoeba.   Digestive enzymes stored within lysosomes
hydrolyze the food molecules of the prey into smaller molecules that can be absorbed
through the vacuolar membrane into the cytoplasm.

pass through the vacuolar membrane into the rest of the cell. Indigestible parts
are eventually discharged to the outside. Although digestion in the amoeba can
properly be described as intracellular, we should keep in mind the definite membrane that persists between the material in its food vacuole and the remainder
of the cytoplasm.

It is obvious that phagocytosis can occur only if the food materials available
to the organism are smaller than its phagocytic cells. It is not surprising, therefore, that feeding by phagocytosis is restricted to those organisms that are
adapted to secure food materials much smaller than themselves.

5-3 EXTRACELLULAR DIGESTION

A second solution to the problem of digesting food is to secrete the digestive
enzymes from the cell and digest the food outside the cell, that is, extracellularly.
Once the food is digested, the small, soluble molecules produced (e.g., sugars,
amino acids) can pass by diffusion or active transport across the cell membrane
and into the cell.

Perhaps the simplest approach to extracellular digestion is that employed
by **saprophytes**. Saprophytes secure their food from nonliving but organic matter, such as dead bodies of plants and animals, food products, excrement, etc.
The nutrition of the common bread mold *Rhizopus nigricans* is typical of the
group. It thrives on a piece of moist bread kept in a dark location (Fig. 5-2). The
bread, a man-made product of a once-living wheat plant, supplies all the dietary
needs of the mold. The starch molecules in the bread are too large, however, to
pass directly through the cell membrane. To convert these large, insoluble

Fig. 5–2

A common bread mold, *Rhizopus nigricans,* growing on a piece of bread. This saprophyte secures all its nourishment by secreting digestive enzymes on the bread and absorbing the products of digestion.

starch molecules into smaller, soluble molecules that can enter the cytoplasm requires a starch-digesting enzyme, or **amylase.** *Rhizopus* secretes this enzyme onto the bread and thus digestion occurs extracellularly. The sugar molecules produced are then absorbed into the cytoplasm. This pattern of extracellular digestion of foods is typical of all fungi and most bacteria.

Most animals, too, digest their food extracellularly. Rarely, though, do they live in a location where they are literally surrounded by organic matter. Instead, they secure by one means or another food materials which appear from time to time in their surroundings. They place this food in a pouch or tube within their body, a process called **ingestion.** Then they secrete their enzymes there and digestion takes place. Thus the enzymes are localized where the food is, rather than being secreted freely into the surroundings.

Very little energy is liberated during the process of digesting, i.e. hydrolyzing, macromolecules. Most of the chemical energy stored in starches, proteins, and fats is still locked up in the end products of their digestion: sugars, amino acids, fatty acids, and glycerol. The process of digestion does accomplish a marked reduction in the size of the molecules, however, so that they may be absorbed readily from the ECF into the cytoplasm of cells.

What is the fate of these small, organic molecules once they enter the cytoplasm? In general, two alternatives are available (Fig. 5–3). They may serve as the building blocks for the synthesis of more complex substances. Polysaccharides, lipids, nucleic acids, and proteins are all essential cell components which are synthesized within the cell from their units that have been absorbed from the ECF. This phase of metabolism in which larger, more complex molecules are built up from smaller, simpler ones is called **anabolism.**

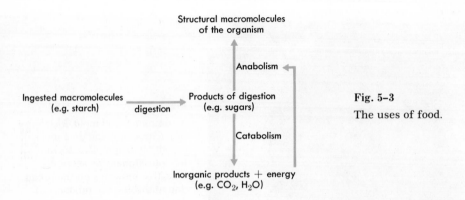

Fig. 5–3
The uses of food.

Many of the digested food molecules absorbed by the cell do not participate in the synthesis of more complex substances, but are instead broken down chemically into still smaller molecules. This breakdown may continue until only simple, inorganic molecules (e.g. H_2O, CO_2, NH_3) remain. The total amount of energy stored in the end products of these decompositions is much less than the amount of energy present in the original molecule. Thus, at each step in the breakdown of sugars, amino acids, fatty acids, and glycerol, energy is released. This phase of metabolism in which relatively complex, energy-rich molecules (e.g. sugars) are broken down into simpler, energy-poor molecules (e.g. CO_2 and H_2O) is called **catabolism**.

ANABOLISM

5–4 REQUIREMENTS

The synthesis within the cell of proteins, nucleic acids, starches, and lipids from the end products of the digestion of these substances requires two additional factors: *energy* and one or more specific *enzymes*.

Energy is necessary as most of these synthetic reactions are energy-consuming. More complex molecules are being constructed from simpler ones and these constructive activities within the cell usually require energy for their accomplishment. The chemical syntheses within the cell of a gland such as the pancreas may require as much energy as a muscle cell uses in violent physical activity. Later in this chapter we will examine the specific mechanisms by which energy is provided for the anabolic, synthetic activities of the cell.

All chemical transformations within cells require specific enzymes in order to proceed rapidly and efficiently. The chemical changes involved in anabolism

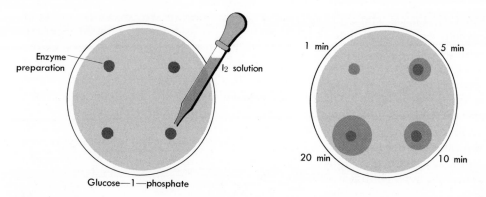

Enzyme preparation

I₂ solution

Glucose—1—phosphate

1 min   5 min

20 min   10 min

Fig. 5–4

Testing the action of the enzyme starch phosphorylase on glucose-1-phosphate incorporated in a dish of agar-agar.   The dark color produced with iodine reveals the progressive synthesis of starch.

are no exception to this rule. In a few cases, the enzymes involved in synthesis may be the same as those involved in digestion. Theoretically, the action of enzymes is reversible. Whether a given reaction proceeds forwards or backwards depends upon other factors such as the relative concentration of reactants and products (see Section 1–12). For example, when one widely occurring enzyme, starch phosphorylase, is placed in contact with a high concentration of "activated" glucose, glucose-1-phosphate, starch synthesis results (Fig. 5–4). (The mechanism by which extra energy is incorporated in the glucose molecule will be discussed later in the chapter.) On the other hand, when the phosphorylase is placed in contact with a high concentration of starch, the breakdown of the starch into units of glucose-1-phosphate is promoted.

In other cases, the enzymes that catalyze the synthesis of large polymers such as proteins and starches are not the same as those involved in the digestion of these substances. In protein synthesis, especially, it is easy to understand why this is so. Man ingests protein from many sources (e.g. beef and beans) and digests these into their constituent amino acids. Once these amino acids are absorbed by his cells, they may be resynthesized into protein. The new protein is not beef or bean protein, however, but human protein. The sequence of amino acids in the protein chains, along with the folding of these chains, results in a protein unique to the species and, unless he happens to have an identical twin, perhaps unique to the individual. It is no wonder then that the enzymes that catalyze the synthesis of a specific protein are quite different from those that catalyze its digestion.

Another important point about the anabolic activities of the cell is that there may be considerable interconversions among the starting materials. For example, excess quantities of ingested glucose may be partially broken down, and then transformed into fatty acids and glycerol. These then can be synthesized into fats. People who get "fat" usually do so as a result of an excessive intake of carbohydrate rather than of fat. Nucleic acids may be synthesized from the absorbed products of nucleic acid digestion, or amino acids can serve as the starting materials instead. In the second case, a longer series of chemical transformations is required. Figure 5–5 illustrates some of the interconversions of the basic building blocks that are possible in the cell. Note that only plants and certain microorganisms can manufacture amino acids from fatty acids plus an inorganic supply of nitrogen (e.g. NH_4^+, NO_3^-). Man, however, can manufacture about a dozen of his required amino acids from other amino acids. The amino acids that he cannot manufacture this way are the "essential" amino acids and must be ingested in his diet.

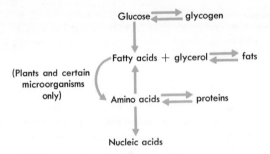

Fig. 5–5
Pathways of interconversions between foodstuffs.

The study of the anabolic activities of the cell is a very active field of research in biochemistry. Our understanding of the mechanisms and pathways involved is still only fragmentary. It is perfectly clear, though, that it is only through these activities that the amount of living matter increases. Growth of organisms requires an increase in the number of cells of which they are composed, or an increase in the size of the cells, or both. In each case, the quantity of those materials intimately associated with life (proteins, nucleic acids, polysaccharides, and lipids) must increase. This increase comes about as a result of the anabolism of the cell. Repair of damaged parts likewise depends upon the synthesis of these large molecules. And, of course, the production of new individuals in the process of reproduction depends directly upon anabolism.

Besides a supply of building blocks and the presence of the necessary enzymes, anabolism requires energy. Let us now examine the mechanisms by which energy is made available within the cell.

CATABOLISM

The catabolic activities of the cell are those in which the organic molecules absorbed by the cell (sugars, amino acids, fatty acids, and glycerol) are broken down into still simpler molecules. In the process of breakdown, some of the energy stored in the starting materials is released. Living organisms accomplish the breakdown by two different (but related) processes: fermentation and cellular respiration.

5–5 FERMENTATION

One method of securing energy from complex organic molecules is to fragment them into simpler molecules. An example of this process is the chemical breakdown of glucose into **ethyl alcohol** and **carbon dioxide.** This process is called alcoholic fermentation. It may be represented by the equation:

$$C_6H_{12}O_6 \rightarrow 2C_2H_5OH + 2CO_2.$$

Inspection of this equation shows that glucose is the only starting material. No other substance, not even oxygen, is required. Because oxygen is not required, the process is **anaerobic;** that is, it can go on in the absence of air. Every atom originally present in the glucose is accounted for in the products. The total energy stored in the products is, however, somewhat less than that stored in the glucose molecule. The energy difference is liberated in the process.

Although the process of alcoholic fermentation has probably been used by man through most of human history, it was not until about 100 years ago that scientists began to understand its nature. The great French biochemist Louis Pasteur found that alcoholic fermentation was always associated with the presence and growth of single-celled fungi, the yeasts. The converse was not true, however. He found that yeasts when exposed to adequate amounts of air produced mostly H_2O and CO_2 as the end products of their metabolism. Pasteur concluded that fermentation was an energy-producing mechanism used by organisms deprived of air. He called fermentation "life without air." Since Pasteur's time, the enzymes and coenzymes that carry out the chemical changes in alcoholic fermentation have been isolated and the details of the process determined. Because of the close relationship between these intermediate steps and those involved in the process of cellular respiration, we will defer a discussion of them until that topic is considered (see Section 5–7).

Other kinds of fermentation are also known. For example, when vigorous activity of our muscles outstrips the supply of oxygen available for cellular respiration, the muscles get energy from the fermentation of glucose. (The glucose is, in turn, derived from animal starch, glycogen, which is stored in muscles.)

Fortunately for us, the end product of this fermentation is **lactic acid,** rather than alcohol and CO_2.

Both alcoholic and lactic acid fermentation are inefficient processes. Although the total energy in the products of fermentation is less than that in glucose, it is not much less. Not more than 4% of the energy stored in glucose is made available to the organism during fermentation. Most of the energy originally stored in the glucose molecule is still stored in the end products of ethyl alcohol or lactic acid. (The use of ethyl alcohol as a fuel component in some racing cars and rocket engines should make this clear.)

Not only is fermentation an inefficient method for securing energy, but it is potentially a hazardous one as well. Ethyl alcohol is poisonous in moderate concentrations. The activity of yeast plants ceases when the alcoholic concentration of the medium in which they are growing reaches about 14%. (This sets a limit on the alcohol concentration of undistilled alcoholic beverages such as natural wines.) The accumulation of excessive amounts of lactic acid in human muscles is undesirable because it lowers the pH of the cell contents and ECF. Too high a concentration of lactic acid causes the muscle to become incapable of further action.

Fermentation is a wasteful process. Only a small fraction of the energy stored in the glucose molecule is made available to the organism. Most of it is still retained in the end products of fermentation. Therefore, organisms that rely on fermentation must have large amounts of glucose available to them in order to satisfy their needs. If, however, glucose could be broken down by cells into truly simple, energy-poor molecules (e.g. H_2O and CO_2), a much higher percentage of its energy would be released. Such a breakdown occurs when glucose is oxidized. Oxidation of glucose takes place within the cells of the vast majority of living organisms in the process of cellular respiration. It is carried out in the mitochondria (Section 3–8).

5–6 CELLULAR RESPIRATION

Cellular respiration can be defined as the oxidation of organic compounds that occurs within cells. In Chapter 1, we defined oxidation as the process of removing electrons from a substance. This occurs when oxygen atoms are added to a substance. When coal (carbon) is oxidized, oxygen atoms combine with the carbon atoms to form carbon dioxide: $C + O_2 \rightarrow CO_2$. The affinity of oxygen for the electrons present in the carbon atom is so great that the oxygen atoms "pull" these electrons close to them. This releases energy which, in the form of heat and light, can be used to heat homes, generate steam, etc. Oxygen is such a powerful and common electron acceptor that it has provided the name for the process: oxidation. Many other substances can, however, serve as oxidizing agents.

In cells, the most common type of oxidation involves the removal of hydrogen atoms (each with its electron) from a substance. The primary oxidizing

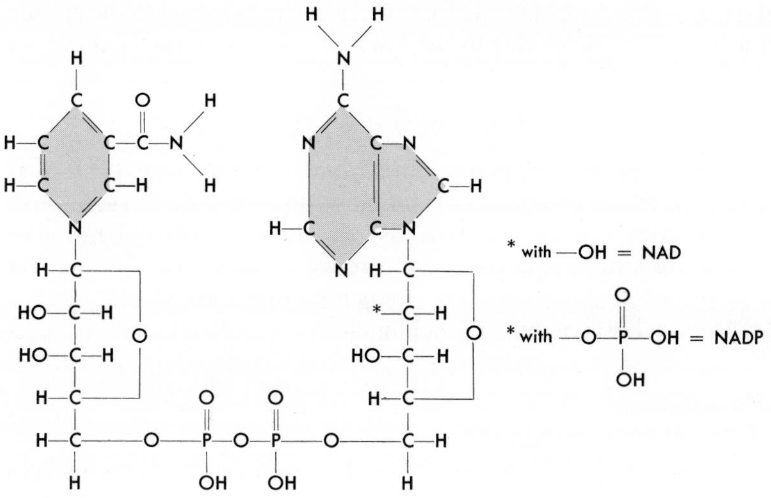

Fig. 5-6

Structural formula of nicotinamide adenine dinucleotide (NAD) and its phosphate (NADP).

agents in cells are the coenzymes nicotinamide-adenine dinucleotide, or **NAD,** and its close relative, nicotinamide-adenine dinucleotide phosphate (NADP). (These two substances have been known by a variety of other names of which DPN, for NAD, and TPN, for NADP, are still quite common.) Each of these is a complex molecule made up of (1) nicotinamide (a B vitamin), (2) two five-carbon ribose units, (3) the purine adenine, and (4) two (in NAD) or three (in NADP) phosphate groups (Fig. 5-6). Both of these substances can remove two electrons (usually associated with two hydrogen atoms) from certain organic compounds, thus oxidizing them. The two electrons actually unite with the NAD (or NADP), thus reducing it. You remember that in Chapter 1 (Section 1-9) we pointed out that the process of reduction is always inseparable from the process of oxidation. An oxidizing agent (e.g. NAD) is reduced by the substance it oxidizes. Because NAD and NADP are so similar in activity, we will refer to either simply as NAD. The reduced form will be indicated as $NADH_2$.

There is another important way in which the oxidation that goes on within cells differs from the burning of coal. The latter process proceeds rapidly with most of the energy being released as heat. The high temperatures involved could not be tolerated by a living organism. The oxidations that occur in living cells proceed rapidly, too, but at low temperatures. Instead of being released as heat, over 40% of the energy liberated in the course of respiration is conserved by the cell as chemical energy. In this form it is available to run the energy-consuming activities of the cell.

The over-all chemistry of cellular respiration has been known for many years. The primary fuel in most living things is the sugar molecule, glucose.

With the aid of oxygen taken in from the ECF, cells oxidize glucose to the simple, energy-poor compounds carbon dioxide and water. The reaction can be expressed by the equation:

$$C_6H_{12}O_6 + 6O_2 \rightarrow 6CO_2 + 6H_2O.$$

Knowing what goes in and what comes out of a process is only part of the story, though. When ignited by a flame, glucose combines directly with oxygen, forming CO_2 and H_2O according to the equation just given. In living cells, however, the process proceeds in a more orderly and controlled fashion. The gradual discovery of the way in which glucose in living cells is oxidized to CO_2 and H_2O has been one of the great achievements of biochemistry.

5-7 Anaerobic Breakdown

Glucose molecules must be activated by heating before they will begin to burn. In cells, also, glucose molecules must be activated before they can be respired. The activation is not accomplished by heat, but instead the energy of the molecule is increased by joining to it a phosphate group,

$$
\begin{array}{c}
\text{OH} \\
| \\
-\text{P}=\text{O.} \\
| \\
\text{OH}
\end{array}
$$

This process is called phosphorylation. Like all chemical reactions in cells, it requires an enzyme. It also requires a source of phosphate. The phosphate is supplied by a substance called adenosine triphosphate, or **ATP** for short. ATP consists of the purine adenine (what other substances contain this molecule?), a five-carbon sugar, ribose, and three phosphate groups (Fig. 5-7). The bonds by which the second and third phosphate groups are attached are known as high-energy bonds because of the large amount of energy stored in them. (They are shown in Fig. 5-7 as wavy lines.) When glucose is phosphorylated, ATP transfers its third phosphate group and the energy of its bond to the glucose molecule. This results in an activated form of glucose, glucose-phosphate. In the same process, the ATP is changed to adenosine *di*phosphate (only two phosphates are left) or **ADP**. A minor enzyme-catalyzed rearrangement of the glucose molecule and the addition of one more phosphate group from a second molecule of ATP results in the formation of fructose-diphosphate. (Fructose is an isomer of glucose; i.e. it has the same molecular formula but a different structural formula.)

The fructose-diphosphate is then split into two halves. The resulting molecules, each containing three carbon atoms, are molecules of phosphoglyceraldehyde or **PGAL**. From this point on, we will trace the path taken by just one molecule of PGAL. Bear in mind, however, that two molecules of PGAL have been produced from the original glucose molecule and hence all reactants and

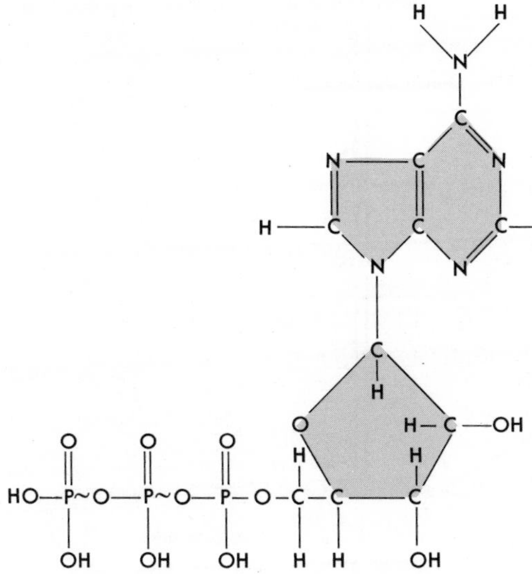

Fig. 5-7

Structural formula of ATP. How does it compare with Fig. 2–14?

products involved from this point on are actually doubled in the process of respiring a single glucose molecule.

PGAL is then oxidized, as two electrons (and their hydrogen nuclei) are removed by NAD. A second phosphate group (donated by a molecule of the inorganic acid phosphoric acid, H_3PO_4) is added and immediately removed and transferred to a molecule of ADP to form ATP. The compound that results from these reactions is phosphoglyceric acid or **PGA** (Fig. 5–8). Two more chemical transformations result in the formation of a second molecule of ATP. With all the phosphate groups removed, a molecule of **pyruvic acid** ($C_3H_4O_3$) remains.

Note that the conversion of glucose to pyruvic acid does not require the presence of oxygen. It is an anaerobic process. Note also that although energy in the form of two ATP molecules is needed to "prime" the process, four ATP molecules (two from each of the PGAL molecules) are produced. There is thus a net production of two molecules of ATP for each glucose molecule used. The total amount of energy stored in two pyruvic acid molecules is less than that stored in the glucose molecule. Some of the difference has been stored in the two molecules of ATP. Some of it is stored in the $NADH_2$. The remainder is liberated as heat.

The energy stored in $NADH_2$ may be used directly by some energy-consuming activity of the cell. One example is the reduction of pyruvic acid to lactic acid:

$$C_3H_4O_3 \rightarrow C_3H_6O_3.$$
$$\text{NADH}_2 \curvearrowright \text{NAD}$$

This reaction is carried on in the cells of some bacteria and in the cells of some muscles. In fact, this reaction is the final step in muscular fermentation (Sec-

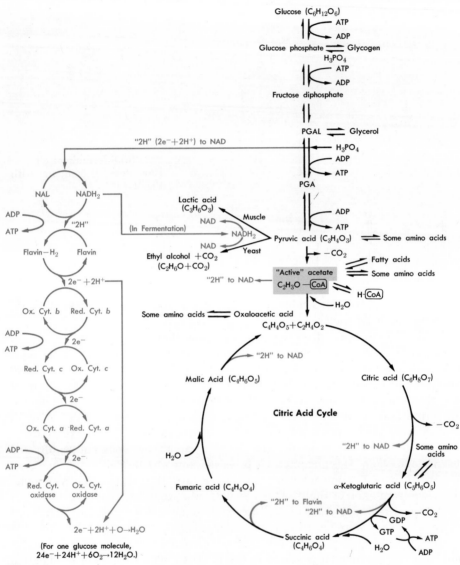

Fig. 5-8

The pathway of cellular respiration.

tion 5–5). Remember that the breakdown of glucose to pyruvic acid is an anaerobic process.

Careful biochemical studies have shown that the anaerobic breakdown of glucose to pyruvic acid proceeds in basically the same step-by-step fashion in both cellular respiration and muscle fermentation. In muscle fermentation, however, the $NADH_2$ formed by the oxidation of PGAL is used up in the reduction of pyruvic acid to lactic acid. There is, however, a net yield of two ATP molecules for every glucose molecule fermented, and it is this net yield which permits the muscle to continue to function for a time even when oxygen is not available.

It is interesting to note that alcoholic fermentation also proceeds by the same biochemical pathway that we have been discussing. In alcoholic fermentation, however, a carbon dioxide molecule is removed from the pyruvic acid before reduction by $NADH_2$ occurs. This reaction thus yields a molecule of CO_2 and a molecule of ethyl alcohol:

$$C_3H_4O_3 \rightarrow C_2H_5OH + CO_2.$$

$$NADH_2 \curvearrowright NAD$$

Again, the organism (e.g. yeast) must satisfy all its energy needs from the net production of two ATP molecules for each glucose molecule consumed. Is it any wonder that yeast cells deprived of air grow relatively slowly and consume large amounts of glucose?

Every step in the anaerobic breakdown of sugar—whether in muscles, yeast cells, or as the first steps in cellular respiration—requires the catalytic action of a specific enzyme. All the enzymes participating in this anaerobic process are present in the cytoplasm of the cell but they have not been found to be localized in any specific cell structures. We shall see, however, that all the remaining steps in the oxidation of sugar are accomplished by enzymes in the mitochondria.

5-8 The Cytochrome Enzyme System

In respiring cells, the usual fate of the $NADH_2$ is to be oxidized in turn by another coenzyme, a yellow substance called a flavin (Fig. 5-8). This oxidation, in some way as yet unknown, is coupled with the production of one ATP molecule from ADP and inorganic phosphate. The electrons received by the flavin are in turn passed on to another enzyme, called cytochrome *b,* thus reducing it. Cytochrome *b* is, in turn, oxidized by cytochrome *c.* This oxidation produces still another molecule of ATP. The electrons then pass to cytochrome *a* and then to the enzyme cytochrome oxidase. This last transfer results in the formation of a third ATP molecule. All these cytochrome enzymes are proteins with iron-containing prosthetic groups (Fig. 5-9). Iron, you remember (see Section 1-9), has a variable valence and thus is admirably suited to serve in REDOX systems. The prosthetic groups of the cytochrome enzymes are also quite similar to the prosthetic group, heme, of the red, oxygen-carrying pigment, hemoglobin, found in vertebrate blood. They also resemble the green plant pigment, chlorophyll, although the chlorophyll molecule contains an atom of magnesium rather than iron.

Cytochrome oxidase donates its two electrons to an atom of oxygen which also picks up two H^+ ions to form a molecule of water. The ultimate electron acceptor then is oxygen just as it is when glucose is burned after being ignited by a flame. In the cell, however, the passage of electrons from the glucose to

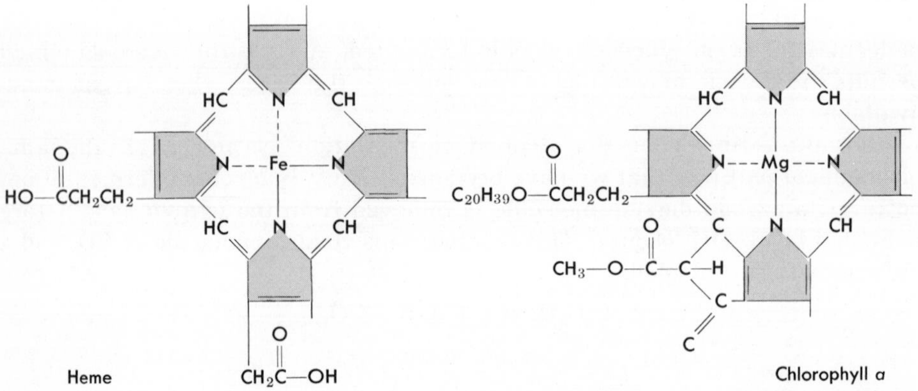

Fig. 5–9

Simplified structural formulas of heme and chlorophyll *a*.

the oxygen proceeds in such small, separate steps, that over 40% of the energy released in the oxidation can be conserved as chemical energy stored in ATP, rather than being liberated as heat.

One might well expect that a system requiring the efficient passage of electrons through a whole sequence of enzymes would have to be highly organized. There is considerable evidence that this is indeed the case. The various cytochrome enzymes have been found to be fixed to the membranes of the cristae of the mitochondria (Fig. 5–10). This discovery was made by rupturing cells and spinning the cell contents at high velocities in a machine called a centrifuge. The centrifugal force developed in these machines causes the various cell components to fall to the bottom of the container, the heaviest ones first, the next heaviest after, and so on. In this way mitochondria have been isolated. The mitochondria can, in turn, be broken by subjecting them to powerful supersonic sound waves. When the outer membrane ruptures, the fluid contents of the mitochondrion are released. However, the membrane itself still possesses the ability to oxidize $NADH_2$ and ultimately to form H_2O. For this reason, we suspect that the various enzymes in the cytochrome system are arranged in an efficient, assembly-line fashion along the membranes of the mitochondrion. The discovery of rows of subunits projecting from the membranes of the cristae lends support to this view (Fig. 5–10). These subunits are thought to contain one or more molecules of enzymes used in cellular respiration.

The burning of 180 gm of glucose (one **mole**) yields 686,000 calories of energy in the form of heat. One mole of ATP stores about 8000 calories of energy. At this point in our story, a net of eight ATP molecules have been produced from one glucose molecule (a net of two from the anaerobic breakdown of glucose and six from the passage of two *pairs* of electrons through the cytochrome system). Only 64,000 calories (less than 10%) of the energy present in the glucose has

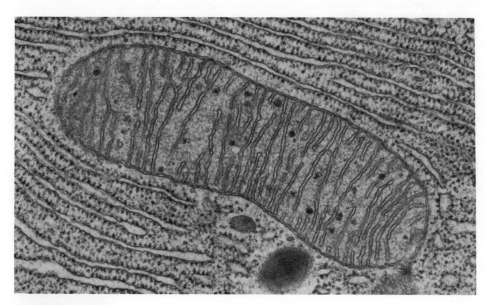

Fig. 5-10

Organization of mitochondria.   Above: The way in which
the inner membranes project into the interior to form the
cristae shows clearly in this mitochondrion from the
pancreas of a bat (70,000 X, courtesy Keith R. Porter).   Left:
The subunits on the surface of the cristae can be seen in the
mouse liver mitochondrion (200,000 X, courtesy Dr.
Donald F. Parsons, *Science* **140**, 985, May 31, 1963).

been extracted. The explanation lies in the fact that pyruvic acid itself is a rea-
sonably complex, energy-rich molecule. The oxidation of pyruvic acid is the next
part of our story.

5-9 The Citric Acid Cycle

To extract the maximum amount of energy from a molecule of glucose, the glucose must be oxidized completely to CO_2 and H_2O. In converting glucose into
two molecules of pyruvic acid, only a partial oxidation has been achieved: four
electrons (and their associated H^+ ions) have been removed from the molecule.

and passed through the cytochrome enzyme system in the mitochondria. In the cellular respiration of glucose, however, the process does not stop with the formation of pyruvic acid. Instead, the molecule is oxidized still further until only the waste materials CO_2 and H_2O are left. This oxidation of pyruvic acid goes on in a step-by-step fashion just as does the breakdown of glucose to pyruvic acid. Each step in the process requires the presence of a specific enzyme. These appear to be localized in or on the membranes of the mitochondria as are the cytochrome enzymes.

The first step in the oxidation of pyruvic acid is the removal of a molecule of CO_2 (Fig. 5–8). The residue is then "activated" by being combined with a coenzyme, called coenzyme A. This complex molecule is called active acetate. In the process of activation, two electrons (with H+'s) are transferred to NAD and from NAD into the cytochrome system. With the aid of a molecule of H_2O, the active acetate complex is broken, coenzyme A is regenerated, and the acetate portion of the molecule ($C_2H_4O_2$) is united with a molecule of oxaloacetic acid ($C_4H_4O_5$). The new molecule ($C_6H_8O_7$) is **citric acid.** (This is the acid which is found in citrus fruits—lemons, grapefruit, etc. It has provided the name for the sequence of chemical changes we are discussing.) After a series of enzyme-catalyzed molecular changes, including removal of another molecule of CO_2 and the removal of another pair of electrons by NAD, alpha-ketoglutaric acid is formed. Another molecule of CO_2 and two more electrons (with H+'s) are removed from the alpha-ketoglutaric acid by NAD. A molecule of H_2O enters the reaction at this step. The result of these changes is succinic acid. The molecular rearrangements involved in the conversion of alpha-ketoglutaric acid to succinic acid also result in the production of one molecule of ATP. The mechanism by which this occurs is not yet clear.

Two more electrons (and H+'s) are removed from succinic acid to form fumaric acid.

Succinic acid   Fumaric acid

The oxidizing agent in this case is flavin. Do you suppose that as many molecules of ATP are produced when flavin donates electrons to the cytochrome system as when $NADH_2$ does? (Figure 5–8 can help you answer this.)

The addition of a molecule of H_2O to fumaric acid produces malic acid. NAD then removes two electrons (and two H+'s) from malic acid with the formation of oxaloacetic acid. What do you suppose then happens to the oxaloacetic acid that has been formed?

In passing through the chemical changes of the citric acid cycle, the active acetate has been completely oxidized to H_2O and CO_2. When the process is completed, the "carrier" molecule, oxaloacetic acid, is regenerated and can then pick up another molecule of active acetate.

5–10 Importance of Cellular Respiration

You may well ask what is the significance of oxidizing glucose by such an involved sequence of chemical reactions. The answer should be clear once we see just how much ATP is produced during the respiration of one molecule of glucose (Fig. 5–8). Remember that a single molecule of glucose gives rise to two molecules of PGAL, whose fate we have just examined. At five different times in the breakdown of one PGAL molecule, NAD removes a pair of electrons and passes them through the cytochrome system. As each pair of electrons passes through the cytochrome system, three molecules of ATP are produced. Hence 15 molecules of ATP are produced from each PGAL molecule by this process, making a total of 30 ATP's for each glucose molecule. In addition one pair of electrons is removed by flavin, but these give rise to only two ATP molecules, since they pass down a shorter length of the chain of enzymes on the mitochondrial membranes. The shortage is made up, however, by the single ATP produced in the transformation of alpha-ketoglutaric acid into succinic acid. Again, as the cycle must "turn" twice for each glucose molecule oxidized, a total of six more ATP molecules are produced by these mechanisms. Adding the net yield of two molecules of ATP which are produced when glucose is broken down anaerobically to pyruvic acid, we end up with a grand total of 38 molecules of ATP produced by the oxidation of a single molecule of glucose.

As mentioned in Section 5–8, a mole of glucose yields 686,000 calories of energy when fully oxidized. In the rapid, uncontrolled oxidation which we call burning, all this energy is liberated as heat and light. The controlled, step-by-step oxidation of glucose in cellular respiration, in contrast, permits a substantial part of this energy to be conserved in the form of chemical energy. This chemical energy, stored in the high-energy phosphate bonds of ATP, can then be used to carry out the many energy-consuming activities of the cell. The efficiency of the process can be calculated easily. The conversion of a mole of ADP to ATP requires about 8000 calories. The respiration of a mole of glucose gives rise to 38 moles of ATP. Thus about 304,000 calories are conserved, or 44% of the energy (686,000 calories) available in the glucose molecules. An efficiency of 44% in the conversion of the chemical energy of glucose into the chemical energy of ATP compares very favorably with the 15–30% efficiency of internal combustion engines in converting chemical energy into mechanical energy.

Not only can we set up a balance sheet to study the energy conversion process, but we can also set up a balance sheet indicating the substances consumed and the substances produced by respiration. Obviously we start with one molecule of glucose. As this is broken down, it gives rise to 12 pairs of electrons (and H^+'s) which pass through the cytochrome enzyme system and unite with 6 molecules of oxygen to form 12 molecules of H_2O. Although only 12 hydrogen atoms and 6 oxygen atoms are present in the original glucose molecule, the deficit is made up by the introduction of 6 (three at each turn of the cycle) water molecules into the citric acid cycle. Six molecules of CO_2 are also produced during these transformations, three from each pyruvic acid molecule entering the cycle. Thus the overall chemical change in cellular respiration can be expressed by the equation:

$$C_6H_{12}O_6 + 6O_2 + 6H_2O \rightarrow 6CO_2 + 12H_2O + 304{,}000 \text{ calories}$$

stored in ATP. See whether you can verify the correctness of this equation by studying Fig. 5–8.

The step-by-step oxidation of glucose into H_2O and CO_2 has another important function. Several of the *intermediate* compounds formed in the process link glucose metabolism with the metabolism of other foodstuffs. Thus PGAL is a breakdown product of glycerol (a component of fats) as well as an intermediate in the breakdown of glucose. Active acetate is produced by the breakdown of the fatty acid component of fats. It is also produced when certain amino acids have their nitrogen removed in the process of **deamination.** Pyruvic acid, alpha-ketoglutaric acid and oxaloacetic acid also may be formed from amino acid breakdown. These links thus permit the oxidation of excess fats and proteins in the diet. No special mechanism of cellular respiration is needed by those animals that depend largely on ingested fats (e.g. many birds) or proteins (e.g. carnivores) for their energy supply. These links between carbohydrate metabolism and the metabolism of fats and proteins are useful in still another way. Fatty acids, glycerol (and then fats), and some amino acids can be *synthesized* using these intermediates. As we have seen, a human can become fat as a result of eating excess amounts of carbohydrates. A knowledge of the links between carbohydrate metabolism and fat metabolism shows us how this is possible (see Fig. 5–8).

The steps in the processes of fermentation and cellular respiration are also interesting because they are found in such a wide variety of living things. While some differences have been found in different organisms, most of the enzymes and intermediate compounds in the processes are common to almost all organisms from bacteria to men.

5–11 HOW THESE DISCOVERIES WERE MADE

You may be puzzled as to how biochemists have been able to discover all the intermediate steps in fermentation and cellular respiration. The discovery has required many years, the work of many men, and the use of a variety of ingenious techniques.

One technique in studying intermediary metabolism is to supply a suspected intermediate substance and see (1) whether it is used up and (2) what new substance begins to accumulate. Sometimes this technique is applied to the intact organism, but usually it is applied to isolated organs, tissues, or even cell extracts. The discovery by Hans Krebs that oxaloacetic acid ($C_4H_4O_5$) added to isolated pigeon breast muscle is converted into citric acid ($C_6H_8O_7$) enabled him to work out the reactions in the citric acid cycle or, as it is sometimes called, Krebs' cycle.

Another technique is to attempt the isolation of a specific enzyme. Grinding up a piece of tissue, such as liver tissue, releases the cell contents. The biochemist can then attempt to extract and purify a single enzyme. If this is accomplished, he can then determine the single substrate upon which that enzyme acts and the substance produced by its action.

The use of enzyme poisons has also provided information on intermediary metabolism. A poison that interferes with cellular respiration may actually affect only a single enzyme in the process. For example, the poisonous effect of malonic acid (see Section 4-8) is due to its inhibition of the enzyme that catalyzes the conversion of succinic acid to fumaric acid in the citric acid cycle. When such a situation occurs, the normal substrate (succinic acid in this case) of the poisoned enzyme accumulates and may be identified. Then if the substance normally *produced* by that enzyme can be discovered, using the techniques mentioned above, this substance can be added to the poisoned system. If respiration begins again, as it does when fumaric acid (Fig. 5-8) is added to tissues poisoned by malonic acid, another link in the chain has been established.

One of the best techniques for studying intermediary metabolism is to introduce into an enzyme system molecules which have been "tagged" with radioactive atoms. The gradual appearance of radioactivity in other chemical substances will indicate the pathway by which the chemical changes have occurred. In the next chapter, we will see how this technique was used to determine one of the important steps in the process of photosynthesis.

Finally, all of these special techniques depend for their success upon the multitude of analytical tests that chemists have devised through years of experimentation and logical reasoning. Some of the techniques for determining the identity of chemical substances and for determining the arrangement of the atoms in their molecules were mentioned in Chapter 1.

5-12 THE USES OF ENERGY

The processes of cellular respiration and fermentation transform chemical energy stored in food molecules into chemical energy stored in the high-energy-phosphate bonds of ATP. ATP, in turn, serves as the immediate source of energy for all the energy-requiring activities of the cell. These fall into several categories.

1. **Mechanical Work.** One of the most important ways in which energy is used by living organisms is to carry on mechanical work. This is especially true of animals. Running, swimming, flying, and climbing are just a few common animal activities in which work is accomplished and energy is consumed. In these cases, the mechanical work is brought about by the contraction of special cells, the muscle cells. The force exerted when muscle cells contract can then be transmitted to organs of locomotion, etc. The energy needed to accomplish muscular contraction is supplied by ATP.

The beating of cilia and flagella is also powered by ATP and accomplishes mechanical work. Such action may propel the cell through the medium as the flagella of *Chlamydomonas* do (Fig. 3–12). Ciliated stationary cells move the medium past them. The cells lining our air passages, for example, are ciliated. Their cilia sweep inhaled dust particles, etc. back up to the throat.

2. **Electrical Work.** Energy is also used by living things to accomplish electrical work. The electric eel is a dramatic example of this. Special electric organs in this fish are capable of converting the energy of ATP into a brief current sufficient to stun an adult human. On a less dramatic (but far more important) scale, the energy of ATP is used to create the voltage which exists between the interior and exterior surfaces of most cell membranes. All nervous activity depends upon this kind of electrical activity.

3. **Active Transport.** A third way in which cells use energy is to counteract the passive forces of diffusion. The active transport of ions or molecules from regions of low concentration to regions of high concentration seems to be closely linked to the production and use of ATP by the cell. Several specific examples of active transport were mentioned in Section 4–5. The periodic filling (and contraction) of the contractile vacuole of the amoeba, for example, requires the energy of ATP.

4. **Bioluminescence.** Many organisms are able to give off light. Although the firefly is a common example, the majority of luminescent organisms are found in the oceans. These include certain marine bacteria, protozoans, crustaceans, mollusks, echinoderms, and even fishes. Biochemists do not yet fully understand the bioluminescent process. One thing that is clear, however, is that energy in the form of ATP is essential for the process to occur.

5. **Heat.** Energy is also a source of heat for living things. The mammals and birds are especially dependent upon internally generated heat. They can maintain a fixed body temperature despite fluctuations in the environmental temperature. Generally, the production of heat occurs simply as a by-product of other energy transformations within the cell. As we have already seen, no energy transformation is 100% efficient. For example, when chemical energy is converted into mechanical energy (as in muscular contraction) a substantial amount (70–80%) of the energy is lost in the form of heat. The involuntary muscular

contraction which we call shivering exploits this inefficiency to prevent our body temperature from falling below normal. Thus although heat cannot do work for an organism, its production may be vital to the organism in cold surroundings.

6. Anabolism. As we have seen, the synthesis of large complex molecules from smaller, simpler ones is an energy-consuming process. From the absorbed products of digestion, cells are able to synthesize proteins, polysaccharides, lipids, and nucleic acids. Energy is needed to accomplish these syntheses. The source of energy here, too, is ATP. You have seen (Section 5–4) that before glucose units can be linked together to form starch, they must first be activated. This activation is accomplished by transferring a phosphate group from ATP to the glucose molecule. A specific enzyme is also necessary. In Chapter 8, you will learn that a similar mechanism is employed by the cell in the synthesis of its proteins. The constituent amino acids must first be activated before they can be assembled into a chain. This activation is also accomplished by ATP and specific enzymes. It is interesting that the activation of each of our 20-odd amino acids requires a separate enzyme, but the single substance ATP supplies the energy requirement for them all. Thus the many anabolic activities of the cell (as well as the other energy-requiring activities discussed above) are all linked to the catabolic, energy-producing activities of the cell by one substance: ATP.

It is as a result of the anabolism of the cell that an increase in the amount of living material (growth) takes place. The process of converting ingested food into more living substance is not 100% efficient, although in some organisms, such as carefully bred varieties of poultry, the percentage of conversion may be remarkably high. Broilers can gain about half a pound of live weight for every pound of food ingested. There are several reasons why one would not expect the rate of conversion to exceed this value by any great amount. First, of course, some of the ingested food must be catabolized in order to provide the energy for the anabolic use of the remainder. Second, although the energy transformations of the cell are remarkably efficient, a sizable part of the energy involved in every chemical change is lost as heat. Third, a great deal of the energy produced in catabolism is used in a multitude of other ways, as we have seen. An adult human may maintain a constant weight for a long period of time, but still is dependent, of course, upon ingested food materials as a source of energy for other activities. It is interesting to note in this connection that the high conversion ratio noted for broilers is achieved only when the animals are reared in relative confinement. The more physical activity carried on by an organism, the smaller will be the percentage of its food actually assimilated, that is, used in growth.

This inability of heterotrophic organisms to convert 100% of their ingested food materials into the constituents of their own cells has other important consequences. All heterotrophic organisms depend ultimately upon autotrophic organisms for their food. Animals that eat plants directly (herbivores) will assimilate a small percentage (generally no more than 10%) of the food materials.

Most of the ingested food will be broken down, as we have seen, into inorganic substances (e.g. CO_2 and H_2O). Animals that feed upon these animals (carnivores) will, in turn, be able to assimilate only a portion of the ingested food and thus just a tiny fraction of the food originally manufactured by the autotroph. Man includes both plants and animals in his diet. (He is an omnivore.) Whenever he consumes meat in his diet, however, he is indulging in the luxury of food whose manufacture has already involved substantial losses of matter and energy back to the environment. It is no accident that a pound of beefsteak is more expensive than a pound of corn. It took many pounds of the latter to produce one pound of the former.

In this chapter we have examined how heterotrophic organisms secure and use the complex, energy-rich, organic molecules which we call food. Whether eaten by herbivore or carnivore, the ultimate source of all these organic molecules is the life activities of autotrophic organisms. Only autotrophs can manufacture organic, energy-rich molecules from inorganic, energy-poor raw materials. How they do this is the topic of the next chapter.

EXERCISES AND PROBLEMS

1 Pound for pound, what food yields the most energy when oxidized?

2 In what ways is alcoholic fermentation similar to cellular respiration?

3 In what ways is it different?

4 Into what forms of energy can the chemical energy of ATP be transformed?

5 What parallels can you draw between the process of bread making and the process of brewing? What differences are there?

6 List the substances in the cell that participate in REDOX reactions.

7 Why is the calorie, a unit of heat energy, also used as a measure of chemical energy?

8 How many moles of oxygen are consumed when two moles of glucose are respired? How many moles of carbon dioxide are produced?

9 What is the ratio of moles of oxygen consumed to moles of CO_2 produced when glucose is respired?

10 What happens to this ratio when fats are used as the fuel in cellular respiration?

11 Why are alcoholic beverages usually brewed in sealed containers?

12 What is the maximum weight of alcohol that yeast cells can produce from 90 gm of glucose?

13 What is the net yield of ATP molecules when glycogen rather than glucose is used to produce the glucose phosphate used in cellular respiration?

REFERENCES

1 Siekevitz, P., "Powerhouse of the Cell," *Scientific American,* Reprint No. 36, July, 1957. Describes the structure of mitochondria and how they are studied.

2 Lehninger, A. L., "Energy Transformation in the Cell," *Scientific American,* Reprint No. 69, May, 1960. How mitochondria convert the energy stored in glucose into ATP.

3 Green, D. E., "The Mitochondrion," *Scientific American,* January, 1964. Attempts to relate the chemical activities of the mitochondrion to its internal construction.

4 McElroy, W. D., *Cellular Physiology and Biochemistry,* 2nd ed., Foundations of Modern Biology Series, Prentice-Hall, 1964. Chapters 4–7 discuss metabolic energy and how it is transformed in fermentation and respiration. Structural formulas are used in the equations.

CHAPTER 6

PHOTOSYNTHESIS: TRAPPING THE SUN'S ENERGY

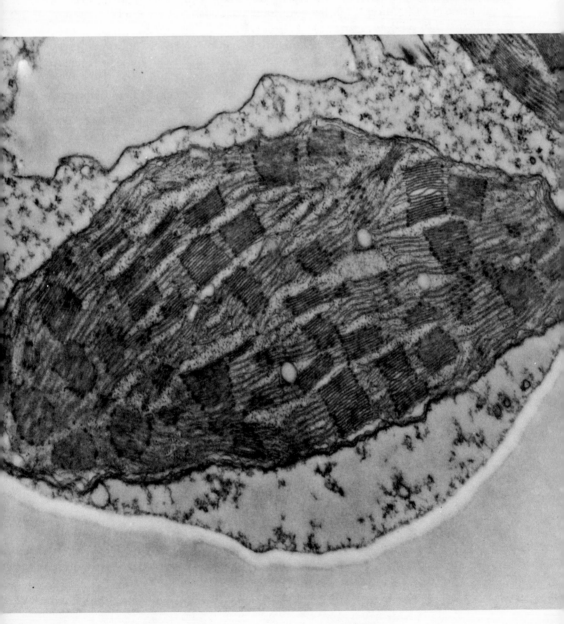

The life of every living organism depends upon a steady supply of materials and energy. The materials are necessary to provide for the growth and repair of the organism. The energy is necessary to permit the organism to maintain its structure in opposition to the tendency of all things (from molecules—see Section 4–3 —to bureau drawers!) to become randomly arranged and disordered. In Chapter 3, we saw how intricately organized is the living cell. Even its constituents, such as the cell membrane, the mitochondria, and the parts of the nucleus, reflect an orderly arrangement of the protein, lipid, and nucleic acid molecules of which they are composed. Cells, in turn, are organized into tissues, the tissues into organs, etc. All these levels of organization depend ultimately upon energy to preserve their pattern. The maintenance of these complex, orderly patterns is one of the main features that distinguish living from nonliving things. With additional energy, additional materials can be organized into these patterns, resulting in growth.

Energy is also necessary for living things to cope with changes in their environment. The activity of nerves and the contraction of muscles permit you to respond to changes in your environment. These activities require energy. Do you think that there is any connection between using energy for this purpose and using energy to preserve the complex organization of matter which we discussed in the last paragraph?

Heterotrophic organisms supply their needs for both matter and energy by taking in complex, energy-rich, organic molecules from their environment. In the last chapter we examined some of the mechanisms by which these are made available to the cells. We also examined the way in which these molecules serve not only as a source of materials for the repair and growth of the cell but also as a source of energy for it.

Where do these complex, energy-rich, organic molecules come from? They are manufactured by green plants and the various kinds of algae. These organisms are capable of synthesizing organic molecules from simple, inorganic materials in the environment such as CO_2 and H_2O. This type of nutrition is referred to as **autotrophic.** Organisms that are capable of autotrophic nutrition not only supply all their own needs for materials and energy but, directly or indirectly, the needs of all heterotrophic organisms. The Bible's statement "All flesh is grass" reflects a crucial biological truth: heterotrophic organisms depend for their existence upon autotrophic organisms. We may dine on beefsteak, but the steer dined on grass.

The kind of autotrophic nutrition upon which almost all heterotrophic organisms (including ourselves) depend is photosynthesis. In photosynthesis, light energy is harnessed for the synthesis of organic compounds from inorganic ones.

◄ Chloroplast (31,000 X). The structures that look like stacks of coins are the chlorophyll-containing grana. (Courtesy Dr. A. E. Vatter.)

6-1 EARLY EXPERIMENTS

Our present knowledge of photosynthesis is the outcome of experiments performed and theories created during the past 300 years. Perhaps the first experiment designed to explore the nature of photosynthesis was that reported by the Dutch physician van Helmont, in 1648. Some years earlier, van Helmont had placed in a large pot exactly 200 pounds of soil that had been thoroughly dried in an oven. Then he moistened the soil with rain water and planted a five-pound willow shoot in it. He then placed the pot in the ground and covered its rim with a perforated iron plate. The perforations permitted water and air to reach the soil but lessened the chance that dirt would be blown into the pot from the outside (Fig. 6-1). For five years, van Helmont kept his plant watered with rain water or distilled water. At the end of that time, he carefully removed the young tree and found that it had gained 164 pounds, 3 ounces. (This figure did not include the weight of the leaves that had been shed during the previous four autumns.)

Fig. 6-1

Van Helmont's experiment. Over a five-year period the willow gained more than 164 lb, while the weight of the soil was practically unchanged.

He then redried the soil and found that it weighed only 2 ounces less than the original 200 pounds. Faced with these experimental facts, van Helmont theorized that the increase in the weight of the willow arose from the water alone. He did not consider the possibility that gases in the air might also be involved.

It was the English chemist Priestley and the Dutch physician Ingen-Housz who discovered that photosynthesis also involves gases in the air. Priestley found that green plants give off oxygen. Thus he was the first to see this important role that plants play in the lives of animals. As we saw in the last chapter, almost all animals require oxygen for cellular respiration. It was Ingen-Housz who, in 1778, showed that the ability of plants to liberate oxygen was entirely dependent upon light. In fact, he found that in the shade or at night, plants consume oxygen, just as animals do. This discovery that plants are capable of respiration plagued later generations of plant physiologists. They could never be sure that their measurements of photosynthetic activity represented true values or simply the difference between an unknown rate of photosynthesis working in one direction and an unknown rate of respiration working in the other. It was not until 1954 that photosynthesis was finally accomplished in isolated chloroplasts and thus free from the presence of mitochondria. Only then could the process be studied independently of the effects of respiration.

That CO_2 plays a part in photosynthesis was first demonstrated (in 1804) by the Swiss chemist de Saussure. He sealed the tops of plants in glass chambers to which he added measured amounts of CO_2. He found that when the plants were exposed to light, they consumed the CO_2 and grew. He also found that they released O_2 at the same rate that they consumed CO_2. By further experiment, de Saussure determined the fate of the carbon atoms in the CO_2. Beans germinated in an atmosphere containing CO_2 grew and increased their total carbon content. Beans germinated in a sealed container free of CO_2 did not increase the amount of carbon present. The total carbon content of the young root, stem, and leaves was no greater than that of the original seed.

6-2 THE PIGMENTS

Ingen-Housz also showed by experiments that only the green portions of plants liberate oxygen. He found that nongreen portions, such as woody stems, roots, flowers, and fruits, actually consume oxygen in the process of respiration. We now know that this is because photosynthesis can go on only in the presence of the green pigment **chlorophyll.** Chlorophyll is a magnesium-containing compound which shows a distinct chemical relationship to the prosthetic group of the cytochrome enzymes (Fig. 5-9). Its chief distinguishing features are the presence of a magnesium (instead of iron) atom and a side chain of carbon atoms called the phytol group. Actually, several kinds of chlorophyll molecules are known. Chlorophyll *a* is found in all green plants and all algae. Chlorophyll *b* is found in plants and green algae along with chlorophyll *a*. Still other chlorophylls are found in other algae and in the photosynthetic bacteria.

Many algae contain other pigments in addition to chlorophylls. These impart the special colors we find in the blue-green algae, the red algae, and the brown algae. They are accessory to the process of photosynthesis.

Chlorophyll is a green pigment because it does not absorb green light. White light (such as sunlight) contains light of all the colors of the visible spectrum from red to violet, but all these colors are not absorbed equally by the pigment. It is possible to determine how effectively each color is absorbed by (1) illuminating a solution of chlorophyll with light of a single color and (2) measuring with a sensitive light meter the amount of light that passes through the solution. By repeating this process with colored lights of the entire visible spectrum, it is possible to draw an **absorption spectrum** (Fig. 6-2). Note that chlorophyll absorbs light most strongly in the red and violet portions of the spectrum. Green light is very poorly absorbed. Hence when white light shines upon chlorophyll-containing structures, such as leaves, green rays are transmitted and reflected, with the result that the structures appear green.

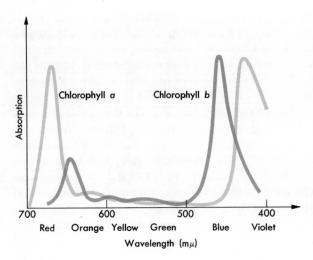

Fig. 6-2

Absorption spectra of chlorophyll *a* and *b*.

Other pigments are found associated with the chlorophylls in green plant cells. Chief among these are the carotenoids. While carotenoids of many colors are known, in plant cells they range in color from yellow to red. Although their presence in green leaves is usually masked by the greater abundance of the chlorophylls, they often become the dominant pigment in flowers and fruits. The red of the ripe tomato and the orange of the carrot are produced by carotenoids. Even in leaves, the carotenoids may become visible. The brilliant yellows of autumn foliage are produced by the carotenoids which become visible as the supply of chlorophyll in the leaf dwindles.

In most green plant cells, the carotenoid content is less than one-quarter that of the chlorophylls. This does not necessarily mean that the carotenoids are not important in photosynthesis. However, the relative importance of the two kinds of pigments can be determined by finding out which colors of light are most effective in causing photosynthesis. In 1881, the German plant physiologist T. W. Engelmann performed a series of experiments to discover this. He placed a filamentous green alga under the microscope and illuminated the strand with a tiny spectrum. In the medium surrounding the strands were motile, aerobic bacteria. After a few minutes of illumination, the bacteria were found to be congregated and most active around the portions of the filament illuminated by red and blue light (Fig. 6-3). Assuming that the bacteria were congregating in the regions where oxygen was being evolved, Engelmann concluded that the red rays and the blue rays are the most effective colors for photosynthesis. By plotting the effectiveness of each color of light in stimulating photosynthesis, one can draw an **action spectrum** (see Fig. 6-3). The similarity of the action spectrum of photosynthesis and the absorption spectrum of chlorophyll suggests that the

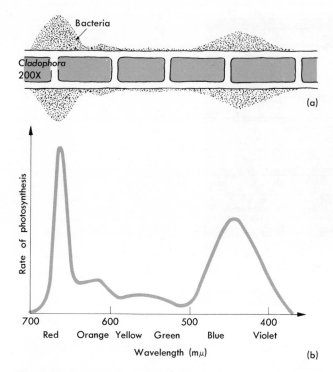

Fig. 6-3

(a) Engelmann's experiment and (b) action spectrum of photosynthesis.

chlorophylls are the most important pigments in the process. The spectra are not identical, however, and this suggests that the carotenoids play some role in photosynthesis.

6-3 CHLOROPLASTS

The chlorophylls (along with some carotenoids) are contained within cell structures called chloroplasts. In the cells of plants, these are disk-shaped. In some of the green algae they may occur in a variety of shapes. For example, the chloroplast in *Chlamydomonas* is cup-shaped (Fig. 3-12). There are no chloroplasts in the blue-green algae. In these organisms, the pigments are confined to membranes arranged concentrically within the cell (look back at the opening illustration in Chapter 3).

In addition to the photosynthetic pigments, chloroplasts contain substantial quantities of lipids and proteins. The latter include many enzymes, among them cytochrome enzymes similar to those found in the mitochondria.

Chloroplasts arise from tiny, colorless structures in the cells called proplastids. Light is necessary for the development of proplastids into chloroplasts.

Fig. 6-4

Seedlings of the common garden bean grown in light (left) and in darkness (right). The pale color of the dark-grown plant is caused by its lack of chlorophyll. When the food reserves of its seed are used up, the seedling will die. Each seedling shows three nodes, but the internodes are greatly elongated in the dark-grown seedling, a condition known as etiolation.

This explains the pale color of seedlings grown in the dark (Fig. 6-4). Proplastids are capable of duplicating themselves. In fact, this is the only way in which additional proplastids are formed. The nuclei of green plant cells cannot manufacture them. Thus whenever green plant cells undergo mitosis, it is important that both daughter cells receive proplastids in the cytoplasm as well as the normal chromosome content of the nucleus. It is probably significant that these self-duplicating structures also contain some of the hereditary material, DNA.

The disk-shaped chloroplasts of higher plants average about 7 micra in diameter. Under the light microscope, these tiny bodies appear to be perfectly uniform in structure. With the aid of the electron microscope, however, it has been found that the chloroplast itself is a highly organized structure. The chlorophyll (and accessory pigments such as the carotenoids) are confined in flat layers, the **lamellae**. Viewed in cross section (Fig. 6-5), these lamellae seem to consist of pairs of modified unit membranes with chlorophyll and carotenoids as well as lipid molecules sandwiched between the layers of protein. Viewed on

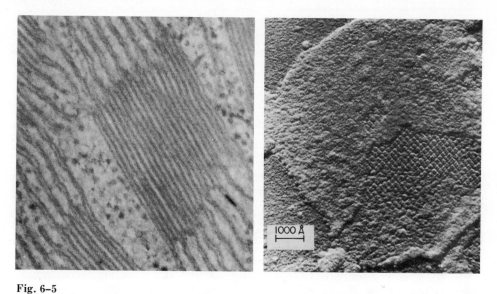

Fig. 6–5

Organization of a chloroplast. Left: Under high magnification (60,000 X), the stacks of lamellae that form the grana can be seen clearly. (Courtesy Dr. A. E. Vatter.) Right: Lamellae are thought to be modified unit membranes. When viewed on their surface, the membranes of the lamellae can be seen to be composed of subunits, the quantasomes. (Courtesy Dr. R. B. Park. From R. B. Park and J. Biggins, *Science* 144, 1009–1011, May 22, 1964.)

their flat surface, the pigments of the lamellae appear to be organized into distinct particles called **quantasomes**. Stacks of lamellae are called **grana** (Fig. 6–5). Surrounding them is a colorless material, the **stroma**. As mentioned above, the intact chloroplast is capable of carrying out the entire process of photosynthesis.

6–4 FACTORS LIMITING THE RATE OF PHOTOSYNTHESIS

With the discovery that the first product of photosynthesis to accumulate in any substantial amount is glucose, it became possible to establish an equation for the process:

$$6CO_2 + 6H_2O \rightarrow C_6H_{12}O_6 + 6O_2.$$

This equation shows the relationship between the substances used in and produced by photosynthesis. It does not, however, tell us anything about the intermediate steps in the process. That there are intermediate steps was first pointed out by the British plant physiologist F. F. Blackman. You, too, can easily repeat his experiment, using the apparatus shown in Fig. 6–6. The green water plant *Anacharis densa* (available wherever aquarium supplies are sold) is the test

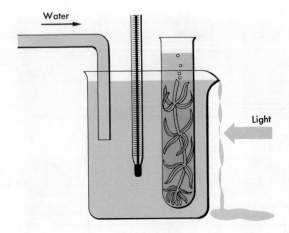

Fig. 6-6

Apparatus for determining the rate of photosynthesis in *Anacharis*. Measurements are made by counting the rate at which bubbles of oxygen are given off at the stem.

organism. When a sprig is placed upside down in a dilute solution of $NaHCO_3$ (which serves as a source of CO_2) and illuminated with a flood lamp, oxygen bubbles are soon given off from the cut portion of the stem. You then count the number of bubbles given off in a definite interval of time at each of several light intensities. Plotting this information should produce a graph similar to that in Fig. 6-7. That the rate of photosynthesis does not continue to increase indefinitely with increased illumination led Blackman to the conclusion that at least two distinct processes are involved: one, a reaction that requires light and the

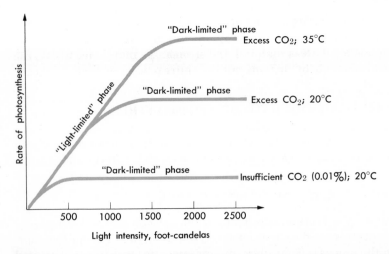

Fig. 6-7

Rate of photosynthesis as a function of light intensity, CO_2 concentration, and temperature. At low light intensities, light is the limiting factor. At higher light intensities, temperature and CO_2 concentration are the limiting factors.

other, a reaction that does not. This latter is called a "dark" reaction although it *can* go on in the light. Blackman theorized that at moderate light intensities, the "light" reaction limits or "paces" the entire process. In other words, at these intensities the dark reaction is capable of handling all the intermediate substances produced by the light reaction. With increasing light intensities, however, a point is eventually reached when the dark reaction is working at maximum capacity. Any further illumination is ineffective and the process reaches a steady rate.

This theory is strengthened by repeating the experiment at a somewhat higher temperature. As we have mentioned, most chemical reactions proceed more rapidly at higher temperatures (up to a point). At 35°C, the rate of photosynthesis does not level off until greater light intensities are present. This suggests that the dark reaction is now working more rapidly. The fact that at low light intensities the rate of photosynthesis at 35°C is no greater than at 20°C also supports the idea that it is a light reaction which is limiting the process in this range. All the light reactions known to chemists depend, not on the temperature, but simply on the intensity of illumination.

The increased rate of photosynthesis with increased temperature does not occur if the supply of CO_2 (in the form of HCO_3^-) is limited. As illustrated in Fig. 6–7, the overall rate of photosynthesis reaches a steady value at lower light intensities if the amount of CO_2 available is limited. Thus CO_2 concentration must be added as a third factor regulating the rate at which photosynthesis occurs. As a practical matter, however, the concentration of CO_2 available to terrestrial plants is simply that found in the atmosphere: 0.03%.

6–5 THE DARK REACTIONS

As you might expect, the dark reaction in photosynthesis is actually a series of reactions. These reactions involve the uptake of CO_2 by the plant and the reduction of CO_2 by hydrogen atoms. Dr. Calvin and his associates at the University of California have devoted years to working out the step-by-step sequence of chemical reactions involved. Their basic experimental procedure has been to expose suspensions of the green alga *Chlorella* to light and to radioactive carbon dioxide. The use of radioactive carbon (C^{14}) in the carbon dioxide "tags" the atom and allows its chemical transformations to be studied. After various intervals of illumination, the *Chlorella* suspension is inactivated and the contents of the cells extracted. These are then separated by a process called paper chromatography. A drop of the cell extract is placed along one edge of a square of absorbent paper. The paper is then dipped into a solvent. The solvent migrates up the sheet because of capillary attraction (Fig 6–8). As it does so, the chemical substances in the drop of cell extract are carried along at different rates. Generally, each compound migrates at a unique rate in a given solvent. When the process is completed, the various substances will be separated at distinct places on the sheet of paper, thus forming what is called a *chromatogram*. The identity

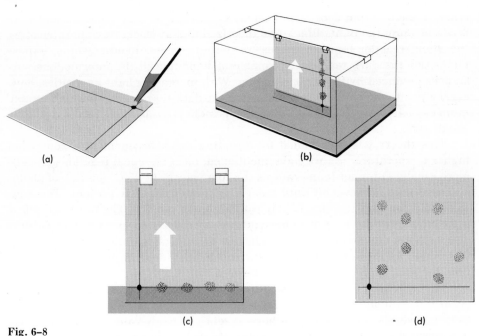

Fig. 6–8

Separation of a mixture by two-dimensional paper chromatography. (a) and (b) Placed in a suitable solvent, the substances in the drop of material at the lower right-hand corner of the paper will be partially separated as they migrate upward. (c) and (d) Further separation can be achieved by turning the paper 90° and using a different solvent.

of each substance may be determined simply by comparing its position with the positions occupied by known substances under the same conditions. Or, the particular portion of the paper can be cut from the sheet and delicate analytical tests run on the tiny amount of substance present.

To determine which, if any, of the substances separated on the chromatogram are radioactive, a sheet of x-ray film is placed next to the chromatogram. If dark spots appear on the film (because of the radiation emitted by the C^{14} atoms), their position can be correlated with the positions of the chemicals on the chromatogram. Using this technique of *radioautography,* Calvin found that C^{14} turned up in glucose molecules within 30 seconds after the start of photosynthesis. When he permitted photosynthesis to proceed for only five seconds, however, he discovered radioactivity in many other, smaller, intermediate molecules. Gradually the pathway of carbon fixation was established.

One of the key substances in this process is the five-carbon sugar ribulose phosphate. When it is "activated" by ATP, the resulting compound, ribulose diphosphate, is capable of combining with CO_2. The resulting six-carbon sugar molecule then breaks down to form two molecules of phosphoglyceric acid (PGA) (Fig. 6–9). Then, in what is fundamentally a reversal of the reaction in cellular

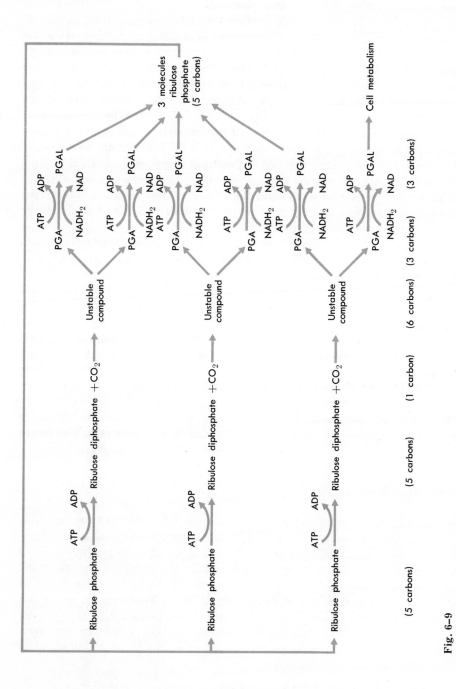

Fig. 6-9
Pathway of carbon fixation in photosynthesis: the "dark" reactions.

respiration (see Section 5-7), the PGA molecule is reduced by $NADH_2$ to form phosphoglyceraldehyde (PGAL). (The energy of ATP is also needed to "drive" this reduction.) Starting with three molecules of ribulose phosphate, six molecules of PGAL are formed. Of these six, five are used to re-form the three starting molecules of ribulose phosphate. The remaining PGAL molecule, whose three carbon atoms represent the net gain from one turn of the cycle, stands right at the crossroads of cell metabolism. It can enter the mitochondria and continue "down" the pathway of cellular respiration. This will lead to the production of energy for the cell and/or the production of intermediates that can be used immediately for amino acid and lipid synthesis. On the other hand, an excess of PGAL will tend to reverse the reactions of cellular respiration. The PGAL then travels "up" the pathway to form fructose and then its isomer, glucose. Glucose and fructose can be combined to form the common table sugar, sucrose. This can then be transported to other regions of the plant. Excess glucose can be stored as the polysaccharide starch.

Hardly had the dark reactions in plant photosynthesis been worked out when it was discovered that these dark reactions are not unique to plants at all. The process of reducing carbon dioxide to form glucose is essentially the reverse of the anaerobic stage of cellular respiration. It has been found to go on in a variety of animal cells, including those of the human liver. However, just as the anaerobic breakdown of glucose produces energy (ATP and $NADH_2$), so its reversal *requires* ATP and $NADH_2$. (As you can see from Fig. 6-9, nine molecules of ATP and six of $NADH_2$ are needed for a net gain of one PGAL molecule.) Animal cells can thus never hope to accumulate glucose by this method. Remember that they must respire glucose to get their ATP and $NADH_2$ and that the process is only 44% efficient.

The discovery that the dark reactions of photosynthesis are not unique to green plants does, however, focus our attention on the light reaction. The secret of autotrophism must lie here.

6-6 THE LIGHT REACTIONS

It was the American microbiologist Van Niel who first glimpsed the role that light plays in photosynthesis. He arrived at his theory through studying photosynthesis in the purple sulfur bacteria. These microorganisms produce glucose from CO_2 as do the green plants, and they need light to accomplish the synthesis. Water, however, is not used as a starting material. Instead, these bacteria use a gas (which is poisonous to us), hydrogen sulfide (H_2S). Furthermore, no oxygen is liberated during this photosynthesis but rather the element sulfur. Van Niel reasoned that the action of light caused a decomposition of H_2S into hydrogen atoms and sulfur. Then, in a series of dark reactions, the hydrogen atoms were used to reduce CO_2 to carbohydrate:

$$CO_2 + 2H_2S \rightarrow (CH_2O) + H_2O + 2S.$$

(The parentheses around CH_2O signify that no specific molecule is being indicated but, instead, the ratio of atoms in some more complex molecule, e.g. glucose, $C_6H_{12}O_6$.)

In these reactions Van Niel envisioned a parallel to the process of photosynthesis in green plants. He reasoned that in green plants the energy of light causes water to break up into H_2 and O. The hydrogen atoms are then used to reduce CO_2 in a series of dark reactions:

$$CO_2 + 2H_2O \rightarrow (CH_2O) + H_2O + O_2.$$

If this theory is correct, it follows that all of the oxygen produced in photosynthesis is derived from water just as all the sulfur produced in bacterial photosynthesis is derived from the H_2S. This conclusion directly contradicts an earlier theory that the oxygen liberated in photosynthesis is derived from the carbon dioxide. If the equation for photosynthesis in Section 6–4 is correct, then at least some of the oxygen released must come from the CO_2. If, however, Van Niel's theory is correct, the equation for photosynthesis would have to be rewritten:

$$6CO_2 + 12H_2O \rightarrow C_6H_{12}O_6 + 6H_2O + 6O_2.$$

Faced with conflicting theories of this sort, scientists try to devise new experiments to test them. The usual method of procedure is to deduce from each theory a "fact" which must be so if the theory is sound. Then experiments must be devised to see which of the deduced "facts," if any, is actually a fact. In this case, the crucial experiments to test the two theories had to await the time when the growth of atomic research made the isotope O^{18} available for general scientific use. You remember that isotopes are atoms whose weight differs but whose chemical properties do not. O^{18} does not happen to be radioactive, but its presence in a compound can be measured. In 1941, a group of biochemists at the University of California allowed *Chlorella* cells to carry on photosynthesis (1) in the presence of water which had been "tagged" with O^{18} and (2) in the presence of CO_2 which had been tagged with O^{18}. If the earlier theory was correct, they would have expected the O^{18} atoms to appear in the liberated oxygen gas in the second experiment but not in the first. If Van Niel's theory was correct, however, the O^{18} atoms should appear in the liberated oxygen in the first experiment but not in the second. The latter proved to be the case. Therefore, we now express the process of photosynthesis by the equation:

$$6CO_2 + 12H_2O \rightarrow C_6H_{12}O_6 + 6O_2 + 6H_2O.$$

These experiments lent great support to Van Niel's idea that the role played by light in photosynthesis was the splitting of H_2O into hydrogen atoms and oxygen atoms. They gave no clue, however, as to the mechanism by which the hydrogen atoms then reduced the carbon dioxide to glucose.

As you have seen (Section 6–5), it was through the work of Calvin and others that this process and the other steps in the sequence of dark reactions were worked out. You also remember that in the dark reactions, the reducing agent is not hydrogen itself, but $NADH_2$. In 1951 it was discovered that chloroplasts illuminated by light reduce NAD to $NADH_2$, which can then be used as the reducing agent in the dark reactions. But there still remained the problem of determining how $NADH_2$ could be manufactured from the hydrogen atoms liberated from H_2O.

The problem has yet to be completely solved. However, Fig. 6–10 illustrates a mechanism which has received considerable experimental support. To understand this mechanism, it is necessary to understand the effect which light has upon chlorophyll.

When a solution of chlorophyll is placed in a beam of white light, it gives off light of a deep red hue. This phenomenon is called fluorescence. It can be demonstrated easily. A crude chlorophyll extract can be prepared by soaking grass leaves in ethyl alcohol. When placed in a beam of white light, the solution fluoresces.

The explanation of the phenomenon of fluorescence is that the energy of the absorbed light is transferred to an electron in the chlorophyll molecule. The electron is "excited" and raised to a higher energy level. In the chlorophyll solution, the electron drops back to its former energy level after a fleeting instant. In so doing, it gives up most of the energy that had excited it in the first place. The energy is released as red light.

When *intact* chloroplasts are illuminated, no fluorescence is observed. This could mean that the excited electrons are removed from the chlorophyll by some other substance before they can drop back. This other substance is called ferredoxin. The excited electrons picked up by ferredoxin are sufficiently energetic to transfer to a molecule of NAD, thus reducing it. H^+ ions, present at all times because of the dissociation of water molecules, can then join with the electrons to give $NADH_2$. Finally, $NADH_2$ transfers its hydrogens to PGA in a dark reaction to form PGAL (Fig. 6–9).

You remember that plants and green algae have two kinds of chlorophyll in their chloroplasts. Chlorophyll *a* appears to play the dominant role in the reduction of ferredoxin. Other pigments may also participate in the trapping of light energy for this process, however; the complex of these pigments is called System I.

To maintain the process of $NADH_2$ production, fresh electrons must be supplied to the chlorophyll molecules in System I to replace the electrons expelled in the reduction of ferredoxin. These fresh electrons come from OH^- ions. Note that in the dissociation of a water molecule, for every H^+ that joins with NAD, an OH^- is left behind (Fig. 6–10). The electron that gives the OH^- its charge is held tightly, and its removal requires a substantial amount of energy. The source of this energy, like that for the reduction of ferredoxin, is light. Light absorbed by pigments in System II causes the chlorophyll molecules to lose electrons and thus become positively charged. In this condition, they are a sufficiently powerful oxidizing agent to remove electrons from the OH^- ions. For every four OH^-

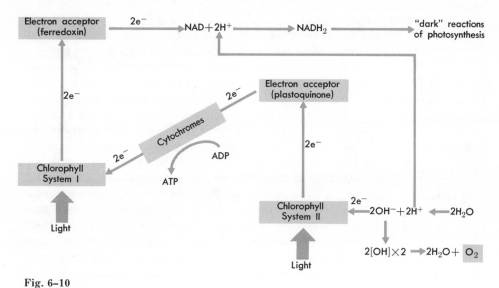

Fig. 6-10

Noncyclic photophosphorylation. The most important light-absorbing pigment in System I is chlorophyll *a*. In System II, most of the light is trapped by chlorophyll *b*. In the red, brown, and blue-green algae, the pigments responsible for their respective colors take the place of chlorophyll *b* in System II.

ions which lose their electrons (become oxidized), two molecules of water and one molecule of oxygen are formed. The electrons are, in turn, raised to a high energy level as light continues to be absorbed by the pigments in System II.

Although chlorophyll *a* is probably responsible for the energizing and transport of electrons in both pigment systems, in System II it acquires much of its energy by transfer from chlorophyll *b* rather than by its own absorption of light. The difference in the absorption spectra of chlorophyll *a* and *b* (Fig. 6-2) increases the range of wavelengths useful in photosynthesis.

Chlorophyll *b* is found only in plants and green algae. The red, brown, and blue-green algae have chlorophyll *a* plus other pigments which give them their respective colors. These other pigments are used in System II, and by transferring their absorbed energy to chlorophyll *a* they markedly extend the range of wavelengths that can be used for photosynthesis by these algae.

The pathway of electrons from System II to System I is not yet entirely clear. At least part of the journey, however, involves cytochrome enzymes in the chloroplasts. As the electrons pass through these cytochrome enzymes, some ATP is manufactured. Then the electrons are further energized during the absorption of light by System I and $NADH_2$ is formed. This process is called **noncyclic photophosphorylation.** It is noncyclic because electrons must be continually supplied by the decomposition of water. It is a photophosphorylation because light energy results in the production of ATP. Clearly, the pigments of System I and II as well as the cytochrome enzymes and the special electron acceptors must be properly oriented with respect to one another so that the flow of electrons can proceed with maximum efficiency. Although the details have yet to be worked out, it is likely that these molecules are precisely arranged in the lamellae of the chloroplasts. Perhaps each **quantasome** (see Section 6-3) contains a single set of the molecules necessary to carry out all the steps in the light reactions.

At first glance, it would seem as though noncyclic photophosphorylation accounted for all the substances needed to convert CO_2 to glucose, namely $NADH_2$ and ATP. Probably not enough ATP is produced by this method, however, to satisfy the needs of all the various dark reactions in photosynthesis.

One possibility for making up the deficit would be to use some of the $NADH_2$ as a reducing agent for the cytochromes in the mitochondria. You remember (Section 5-8) that in cellular respiration, $NADH_2$ donates its electrons (and hydrogen ions) to the cytochrome enzyme system, which eventually passes them on to oxygen. In the process, ATP is manufactured. This possibility seems somewhat unlikely, though, when we see that most green plant cells have a relatively small number of mitochondria compared with the number of chloroplasts they contain.

This possibility was eliminated in 1954 when Dr. Daniel Arnon and his associates at the University of California showed that illuminated chloroplasts can manufacture substantial quantities of ATP even when isolated from other cell components (including mitochondria). Furthermore, this can be done even when no CO_2 is supplied to the chloroplasts and hence no dark reactions can occur. Even when the production of $NADH_2$ itself is stopped (by depriving the chloroplasts of NAD), the chloroplasts continue to produce large quantities of ATP as long as they are supplied with ADP, inorganic phosphate, and light. The energy of light is thus being converted directly into the chemical energy of ATP.

Dr. Arnon named this process **cyclic photophosphorylation.** A possible mechanism is shown in Fig. 6-11. Light absorbed by chlorophyll *a* energizes and expels electrons from the molecule so that they can be trapped by ferredoxin. This then passes the electrons to the cytochrome enzymes in the chloroplast. As they pass "down" the system, their energy is used to produce ATP (just as in the cytochrome enzymes of the mitochondria). The cytochromes then return the electrons to the chlorophyll molecule where they can once again be energized by absorbed light and passed through the cycle.

The process is truly cyclic because no outside source of electrons is required. The chlorophyll is simply trapping light energy and using it to energize its electrons sufficiently to transfer them to the cytochrome enzymes in the chloroplast. On their route back to the chlorophyll molecule, they give up their acquired energy to form ATP. Thus chloroplasts can produce ATP as long as light shines upon them. This remarkable mechanism for converting light energy directly into the chemical energy of ATP more than makes up for the scanty production of ATP during the production of $NADH_2$ in noncyclic photophosphorylation. In fact, cyclic photophosphorylation probably provides (during daylight) the energy

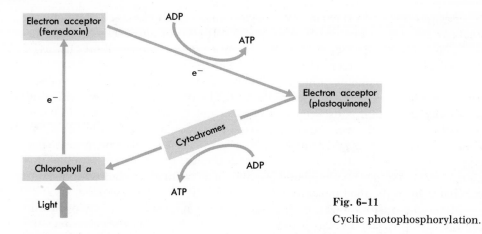

Fig. 6-11

Cyclic photophosphorylation.

for many other energy-consuming reactions of the green plant cell. Not only the dark reactions of photosynthesis, but the energy for starch synthesis, protein synthesis, etc., can be supplied by the ATP produced in cyclic photophosphorylation.

In several respects, cyclic photophosphorylation resembles the second stage of noncyclic photophosphorylation, that is, the stage energized by the pigments in System I. Furthermore, there is some evidence that System II can absorb enough light to carry out noncyclic photophosphorylation without the help of System I. If this turns out to be the case, the role of System I will be simply the manufacture of ATP by cyclic photophosphorylation.

In any case, these light reactions (both noncyclic and cyclic) are thus the key to the ability of green plants to manufacture complex, energy-rich molecules from simple, energy-poor ones. As we have seen, heterotrophic organisms (such as ourselves) fulfill their energy requirements by converting the chemical energy of the glucose molecule into the chemical energy of ATP. The photosynthetic organisms can take care of their energy requirements by converting directly the energy of light into the chemical energy of ATP. In so doing, they produce organic molecules to supply themselves with energy in times of darkness. In producing a surplus of these organic molecules, they make possible the existence of all the heterotrophic organisms on this planet.

In the course of studying the light reactions and dark reactions in photosynthesis, methods were found to disrupt the chloroplasts and separate the pigment-containing lamellae from the stroma. Although the isolated lamellae retained the ability to carry on both cyclic and noncyclic photophosphorylation, they could not convert CO_2 into carbohydrate. These dark reactions were readily carried out by the colorless stroma, however, when it was supplied with CO_2, $NADH_2$, and ATP.

As we saw in the case of the mitochondria, the efficiency of cell reactions cannot be properly understood without considering the physical arrangement of the enzymes that carry them out. When you consider that the green plant chloroplast can manufacture glucose from CO_2 and H_2O after only 30 seconds of illumination, you can appreciate that the many enzymes and pigments involved must be carefully arranged with respect to one another.

In many ways, photosynthesis and cellular respiration are complementary processes. The transport of electrons lies at the heart of each. In photosynthesis, the energy of the sun is used to remove electrons from oxygen atoms in water molecules. These electrons are transferred to carbon atoms, forming covalent bonds between carbon and hydrogen. Carbon is far less electronegative than oxygen and, consequently, energy is needed to make the transfer. Much of this energy is stored in the covalent bonds established. In this way photosynthesizing organisms form the food molecules upon which all organisms depend.

In cellular respiration, the electrons in food molecules are removed and allowed to return, step by step, to oxygen atoms. As they do so, they give back the energy originally stored by them, a substantial fraction of which is converted into the energy-rich bonds of ATP. In the last analysis, then, energy flow in all living things depends on the cyclic transfer of electrons between oxygen atoms and carbon atoms, accompanied by the breakdown (in photosynthesis) and synthesis (in cellular respiration) of water molecules.

EXERCISES AND PROBLEMS

1   The oxygen released in photosynthesis comes from which of the two raw materials used by the plant?

2   When green algae are illuminated simultaneously by two beams of single-wavelength light (about 45 mμ apart at the red end of the spectrum), the amount of photosynthesis that takes place is markedly greater than when just a single wavelength of the same total intensity is used. How does this support Arnon's hypothesis? *Hint:* Refer to Fig. 6–2.

3   Life can be said to be fundamentally a matter of the transport of electrons between energy levels. Defend this view.

4   Distinguish between an autotroph and a heterotroph.

5   What is the most important reducing agent in living cells?

6   What color of light do green leaves absorb least well?

7   Do you think that the rate of photosynthesis would continue to increase indefinitely with increasing temperature? Explain.

8   Do you think that the rate of photosynthesis would continue to increase indefinitely with increasing carbon dioxide concentration? Explain.

9   When students perform the experiment illustrated in Fig. 6–6, starting with minimum light intensity, they often find that the rate of photosynthesis does not level off sharply at higher light intensities. What factor have they neglected to keep constant?

10 Students who have taken an extra-long time to carry out the experiment in Fig. 6–6 often find that the rate of photosynthesis begins to drop at high light intensities. Can you think of an explanation for this?

11 In two columns, contrast photosynthesis and respiration in as many ways as you can.

12 Do green plants need digestive enzymes? Explain.

13 How many electrons are removed from water molecules for each molecule of O_2 produced in photosynthesis?

14 How many electrons are needed for each molecule of CO_2 assimilated in the process of photosynthesis? How many $NADH_2$ molecules? How many ATP molecules?

15 How many ATP molecules are produced when 4 electrons pass from System II to System I? Relate this to the inadequacy of noncyclic photophosphorylation alone to meet the needs of the dark reactions.

REFERENCES

You can read brief firsthand reports of the experiments of van Helmont, Priestley, Ingen-Housz, de Saussure, Engelmann, and van Niel as well as of the O^{18}-tracer experiments done at the University of California in:

1 *Great Experiments in Biology,* ed. by M. L. Gabriel and S. Fogel, Prentice-Hall, Inc., 1955.

Several excellent articles on various aspects of photosynthesis have appeared in *Scientific American.* All of those cited are available in inexpensive reprints:

2 Rabinowitch, E. I., "Photosynthesis," *Scientific American,* Reprint No. 34, August, 1948. A clear introduction to the subject written before the details of the light and dark reactions had been worked out.

3 Wald, G., "Life and Light," *Scientific American,* Reprint No. 61, October, 1959. Discusses the properties of light and the chlorophyll molecule that absorbs it.

4 Arnon, D. I., "The Role of Light in Photosynthesis," *Scientific American,* Reprint No. 75, November, 1960. The light reactions.

5 Bassham, J. A., "The Path of Carbon in Photosynthesis," *Scientific American,* Reprint No. 122, June, 1962. The dark reactions.

6 Lehninger, A. L., "How Cells Transform Energy," *Scientific American,* Reprint No. 91, September, 1961. Emphasizes the reciprocal relationship between photosynthesis and respiration.

7 French, C. S., "Photosynthesis," *This Is Life,* ed. by W. H. Johnson and W. C. Steere, Holt, Rinehart and Winston, New York, 1962. A summary that treats all aspects of the process.

8 Rabinowitch, E. I., and Govindjee, "The Role of Chlorophyll in Photosynthesis," *Scientific American,* Reprint No. 1016, July, 1965.

PART III

CONTROL MECHANISMS
IN THE CELL

Dividing cells as seen with a differential interference
microscope, 650 X. (Research photograph of
Drs. R. D. Allen and A. Bajer, 1966). ►

CHAPTER 7

CELL REPRODUCTION

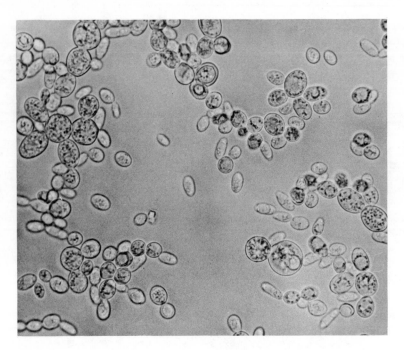

Fig. 7-1

Yeast cells growing in culture. Note the many cells that are in the
process of reproducing.

One of the important achievements of biology in the nineteenth century was the
realization that all living things are composed of cells. Many organisms, such as
yeast cells or a bacterium, consist of simply a single cell. Others, including ourselves, are constructed from many cells. In such *multicellular* organisms, the
cells are of a number of different kinds—blood, nerve, muscle, etc. Each of these
specialized cells performs one or a few specific functions in the organism and
has a structure appropriate to that function. Nerve cells, for example, have
long extensions over which nerve impulses are propagated.

7-1 THE GENETIC CONTINUITY OF CELLS

Shortly after the discovery of the cell as the fundamental structural unit out of
which living things are made, it was recognized that every cell arises from a
preexisting cell. That is, each cell in our body has been formed by a preexisting
cell which, in turn, was formed by another cell. Ultimately, all our cells are derived from a single cell, the fertilized egg with which our life began. Before this,
however, were the sperm cell of our father and the egg cell of our mother. These,
too, were produced from other cells in their bodies.

The central issue of cell reproduction can be seen more clearly in the unicellular organisms. A single yeast cell introduced into a favorable medium will soon give rise to thousands of progeny (Fig. 7-1). And, barring occasional accidents, every one of these offspring will have the same features of structure and function that the first cell had.

This continuity of traits from one generation of cells to the next is nicely exploited in the brewing industry. The flavor of a beer or ale is dependent on a number of factors, one of the most important of which is the particular strain of yeast which is used in the fermentation process. In a typical case, several hundred pounds of yeast cells are added to a vat filled with various ingredients including carbohydrate as an energy source. Because oxygen is excluded, the yeast cells are dependent on fermentation for their energy. As you learned in Chapter 5, fermentation is an inefficient process. This restriction, coupled with the low temperatures used, markedly limits the rate of cell reproduction. Even so, after four or five days, the quantity of yeast in the vat will have increased three- or four-fold. A portion of this population is removed from the mixture and carefully saved to be used to start the next batch of ale or beer. At all times, great care is taken to see that the yeast strain does not become contaminated by other microorganisms. Thanks to such precautions, a single strain of yeast may be used for decades to produce a unique ale or beer. Even with the slow growth rate that takes place under anaerobic conditions, after some twenty years the cells being used are the product of as many as 3000 generations—yet the traits of the original yeast cells have remained unchanged.

Our problem, then, is to account for this self-replicating feature of cells. By what mechanism can the traits of cells be passed unchanged to their offspring through thousands of generations?

7-2 THE CHEMICAL NATURE OF GENES

In 1928, an English bacteriologist, Fred Griffith, made a discovery which, pursued over the next 30 years, ultimately revealed the mechanism of genetic control and continuity in cells. Griffith worked with the bacterium that causes bacterial pneumonia, a widespread and dangerous disease before the discovery of antibiotics. One of the most striking features of this organism is the presence around each cell of a gummy capsule made of a polysaccharide. When these bacteria are grown in culture, the presence of the capsule causes the colonies to have a glistening, smooth appearance (Fig. 7-2). Because of this, the cells are referred to as "S" cells. However, after prolonged cultivation outside the living host, some cells lose the ability to make the capsule. The surface of their colonies then appears wrinkled and rough ("R") (Fig. 7-2). With the loss of the ability to make the capsule, the organisms also loses its virulence. The R forms cannot cause the disease.

Pneumococci, as these bacteria are known, also occur in a large variety of types: I, II, III, etc. Each type produces its own specific polysaccharide

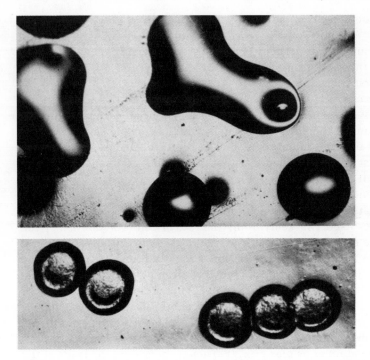

Fig. 7–2

Pneumococci growing as colonies on the surface of a culture medium. Top: the presence of a capsule around the bacterial cells gives their colonies a glistening, smooth (S) appearance. Bottom: pneumococci lacking capsules have produced these rough (R) colonies. (Research photograph of Dr. Harriett Ephrussi-Taylor, courtesy The Rockefeller University and *Scientific American*.)

capsule, which can be distinguished from that produced by other types. Unlike the occasional shift of S → R, the type of the organism is constant. Mice injected with a few cells of one type, say II, will soon have their bodies teeming with descendant cells of the same type. With respect to type, then, we find the same continuity of traits from generation to generation that we found in yeast.

When injected alone, neither living R cells nor S cells that have first been killed by subjecting them to a high temperature cause any symptoms of disease. Griffith unexpectedly found, however, that when living R cells and killed S cells were injected *together* into a mouse, the mouse became ill and *living* S cells could be recovered from its body. In exploring this phenomenon further, Griffith discovered that the type of the S cells was determined by the type of the dead S cells he used, not the type of the strain from which the living R cells were derived. In other words, when a mouse was injected with living R cells from Type I pneumococci and dead S cells of Type II, the living cells that were recovered after the mouse became sick were S cells of Type II. Furthermore, this change of

R-I cells to S-II cells was stable and inheritable. The S-II cells could be cultured indefinitely and remained true to type. Clearly, something in the dead S-II cells had converted the R-I cells into S-II cells, and that something was passed on from generation to generation.

Within a few years, this *transformation,* as it was called, was accomplished in a test tube. A small percentage of R-I cells, grown in test tubes of broth to which *dead* S-II cells were added, became transformed into S-II cells. Later, a cell-free extract of S-II cells was successfully used to transform R-I cells into S-II cells.

As you would expect, this cell extract had a variety of ingredients in it, including polysaccharides, protein, DNA, RNA, and lipids. By selectively destroying one of these ingredients after another and testing the transforming activity of the remaining material, Dr. O. T. Avery and his co-workers were able to show in 1943 that the active ingredient of the transforming principle was DNA. Purified DNA from S-II cells transformed R-I cells into S-II cells. Here, then, was a molecule which changed the function of the recipient cell, making it produce, in this case, a type II rather than a type I capsule. Furthermore, this molecule was self-duplicating. Far more S-II-transforming DNA could be extracted from the offspring of the transformed cells than was used originally to accomplish the transformation. An important question therefore arose: What are the properties of DNA that enable it (1) to dictate the synthesis of Type II polysaccharide and (2) to duplicate itself from generation to generation? The answer to the second part of this question was discovered first.

7–3 THE WATSON-CRICK MODEL OF DNA

As you learned in Chapter 2, DNA is a macromolecule, consisting of repeating units called nucleotides. Each nucleotide consists of three subunits: (1) a sugar group, deoxyribose (hence the name, *d*eoxyribo*n*ucleic *a*cid for DNA), (2) a phosphate group, and (3) a nitrogen-containing base. These bases are generally restricted to four kinds: two purines (adenine and guanine) and two pyrimidines (thymine and cytosine). For convenience, we will refer to these four bases as A, G, T, and C respectively. Furthermore, whatever the quantity of A present in an organism's DNA (and it varies widely from species to species), the amount of T is precisely equal to it. Similarly, the amount of G present is always equal to the amount of C.

Working with this knowledge plus clues as to the shape of the intact molecule and knowledge of the bond angles and shapes of the subunits of DNA, James Watson and Francis Crick proposed a model of the way in which the various constituents of DNA are attached to one another. They deduced that the sugar and phosphate groups alternate with one another in a long strand. Two of these strands are twisted around each other in the form of a double helix, something like a double spiral staircase (Fig. 2–15). A base is attached to each sugar in the chain so that it projects in toward the axis of the staircase. There it joins with the base projecting from the other staircase. Because of the shape of the purine and pyrimidine bases, and the location of their bonding sites, A can be attached only to T. Similarly, G can be attached only to C. This, then, explains why the amounts of A and T (also C and G) in the molecule are always equal. Single DNA molecules may contain from hundreds of thousands to millions of base pairs.

The Watson-Crick model provides a clue to the manner in which hereditary information could be stored in the DNA molecule. Although the basic arrangement of parts in the DNA molecule seems to be the same for all organisms, there is an almost infinite variety of sequences in which these four kinds of base pairs, present by the hundreds of thousands, can be arranged. The International Morse Code provides an interesting parallel. It is made up of only three basic units: a dot, a dash, and a pause. Given sufficient patience and time, however, a telegraph operator could transmit the contents of an entire library by the sequences in which he combined these three units.

At one "step" on a single "staircase" in the DNA molecule, any one of the four bases may be present. Thus there are four items in our code. Note, too, that whatever base is present at a given step, we know immediately what kind of base bridges the gap between it and the opposite staircase. Thus whatever sequence of bases is present on one strand of the DNA molecule, a complementary sequence is present on the other strand. The two sequences have the same relationship to each other as the "positive" and "negative" of a photograph. This redundant nature of the code provides a nice mechanism to explain the property that Avery discovered in DNA: its ability to duplicate itself between one cell generation and the next.

According to the Watson-Crick hypothesis, DNA duplication begins with an "unzipping" of the "parent" molecule; that is, the bonds between the base pairs become broken and the two halves of the molecule unwind. Once exposed, the bases on each of the separated strands can pick up the appropriate nucleotides present in the surrounding medium. Each exposed C will pick up a G, each G a C, etc. With the aid of a linking enzyme, these are polymerized into a complementary chain. When the process is completed, two complete DNA molecules will be present, identical to each other and to the original molecule (Fig. 7–3).

In 1958, this hypothesis received strong experimental support from the work of M. S. Meselson and F. W. Stahl. They used the common intestinal bacterium *Escherichia coli* as their experimental organism. *E. coli* cells can grow in a culture medium containing simply glucose and inorganic salts. Among the latter must be a source of nitrogen atoms for protein and nucleic acid synthesis. Nitrate ions (NO_3^-) serve nicely. Although the most common isotope of nitrogen is N^{14}, it is possible to synthesize nitrates containing a heavier isotope of nitrogen, N^{15}. Meselson and Stahl first grew *E. coli* cells for several generations in a medium containing $N^{15}O_3^-$. They found that at the end of this period the DNA of the cells was about 1% heavier than normal because of the incorporation of N^{15} atoms in it. Then they transferred the cells to a medium containing ordinary

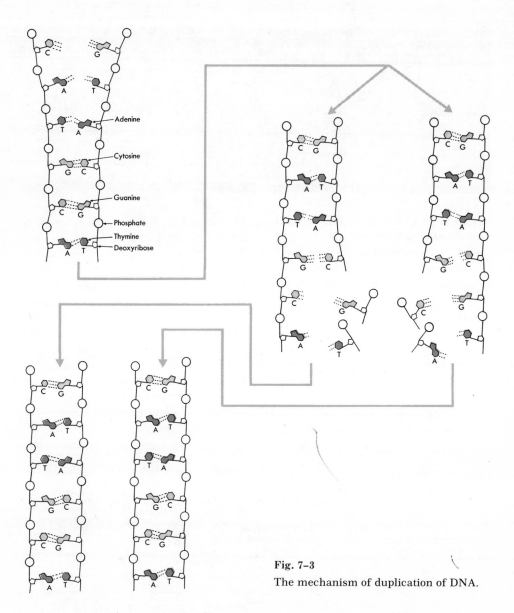

Fig. 7–3

The mechanism of duplication of DNA.

nitrate ($N^{14}O_3^-$) and allowed them to divide just once. The DNA in the new generation of cells was exactly intermediate in weight between the heavier DNA in the previous generation and the normal. This, in itself, is not surprising. It tells us no more than that half the nitrogen atoms in the new DNA are N^{14} and half are N^{15}. It tells us nothing about their arrangement in the molecule. However, when the bacteria were allowed to divide *again* in normal nitrate ($N^{14}O_3^-$), two distinct weights of DNA were formed: half the DNA was of normal weight

and half was of intermediate weight. As shown in Fig. 7–4, this indicates that DNA molecules are not degraded and reformed between cell divisions, but instead, each original strand remains intact as it builds a complementary strand from the nucleotides available to it.

The structure of the DNA molecule not only permits accurate self-duplication but also is ideally suited to the long-term preservation of the sequence of bases coded in it. Because the two chains in a DNA molecule are complementary, the information in one is also coded in the other. In a sense, then, each DNA molecule contains two copies of its information. If for some reason the sequence of bases on one chain were disturbed, the correct message could still be determined from the complementary chain. Recently it has been discovered that the cells of some bacteria and probably even of some higher organisms have mechanisms with which they correct errors that occur in their DNA code. By

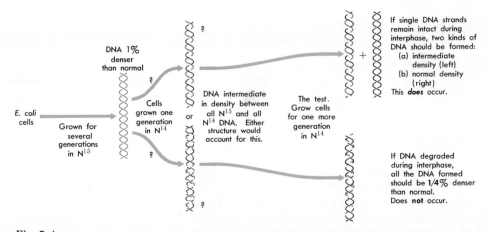

Fig. 7–4

The experiment of Meselson and Stahl. The results (right) suggest that during the process of duplication each strand of DNA remains intact and builds a complementary strand from the nucleotides available. This is consistent with the Watson-Crick model of DNA.

Fig. 7-5

Electron micrograph of a DNA molecule entering a pneumococcus (the large object at the right). This DNA molecule, the long fine line, is approximately 7 microns in length, long enough to include a dozen genes. The process of *transformation* follows the uptake of such a molecule by the bacterium. (38,000 X, courtesy Dr. Alexander Tomasz.)

means of enzymes, they remove damaged or incorrect bases from their DNA molecules. These are then replaced by the correct bases, that is, bases complementary to the bases on the opposite strand. In this way, the genetic code can be preserved intact.

As work with transformation by DNA progressed, it was discovered that, on occasions, two or even three different traits could be transferred simultaneously by the process. If, for example, Type II pneumococci sensitive to the antibiotic penicillin and also sensitive to another antibiotic, streptomycin, were treated with DNA extracted from Type III pneumococci that were resistant to both these antibiotics, most of the transformed cells picked up only one of the traits. However, some picked up two traits and a very few picked up all three. This suggested that a given DNA preparation could carry the information for a number of traits (Fig. 7-5). In the terminology of classical genetics, we can say that these DNA molecules carried several *genes*. In fact, exhaustive studies with *E. coli* have demonstrated that *all* the known genes or this organism are carried on a *single* DNA molecule which is over 1 mm long. (The organism itself is only about 2μ long!)

7-4 MITOSIS

The situation is more complex in higher organisms. Virtually all their DNA is incorporated in the chromosomes. How many DNA molecules are present in each chromosome is not known, nor is it known how the DNA molecules are organized in the chromosome. It is well established, however, that chromosomes, like DNA, do duplicate between cell divisions. When a cell gets ready to divide,

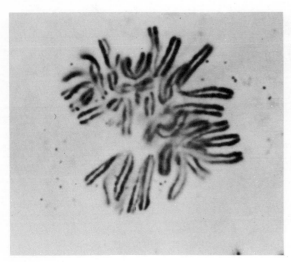

Fig. 7–6

Chromosomes in a dividing epidermal cell of a salamander. The duplicated chromosomes are still attached to each other by centromeres. (Courtesy General Biological Supply House, Inc.)

the chromosomes condense and can be examined under the light microscope. They appear with the duplicates held together by a small body called a centromere (Fig. 7–6). Each centromere attaches to a spindle fiber—an array of microtubules (see Section 3–13) that extends between the ends of the cell. The centromeres move to the center of the cell and duplicate. Each of the two resulting centromeres remains attached to one of the duplicate chromosomes. The centromeres, with their attached chromosome, now move apart, still on the spindle, and migrate to opposite ends of the cell (Fig. 7–7). This process, called mitosis, is thus a device for the orderly separation of the chromosomes that were duplicated prior to the onset of cell division. If indeed the genes are an integral part of chromosomes, then mitosis is a device for parceling out identical sets of genes to the two daughter cells.

Although the way in which DNA is organized in the chromosomes is still unknown, the duplication of the chromosomes, like that of E. coli DNA, appears to be semiconservative, that is, the parental chromosome structure seems to remain intact during the duplication process. This was first demonstrated by Dr. J. Herbert Taylor. He grew plant cells in a solution containing radioactive thymidine (the nucleotide containing thymine, T) until their chromosomes became uniformly radioactive. He then transferred the cells to a medium containing nonradioactive thymidine and an inhibitor of spindle formation. After one duplication in this medium, the chromosomes were still uniformly radioactive. After a second duplication, however, one arm of each duplicate set was radio-

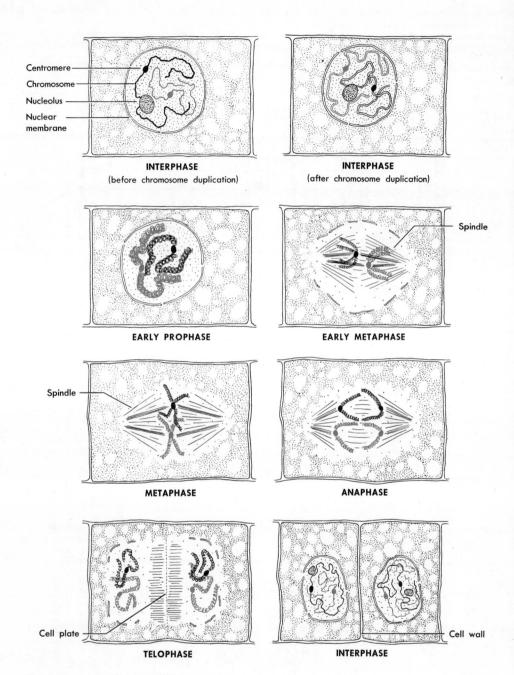

Fig. 7-7

Mitosis in a plant. The stages are shown in semidiagrammatic fashion. For the sake of clarity, only one pair of homologous chromosomes is shown: one member in black, the other in color.

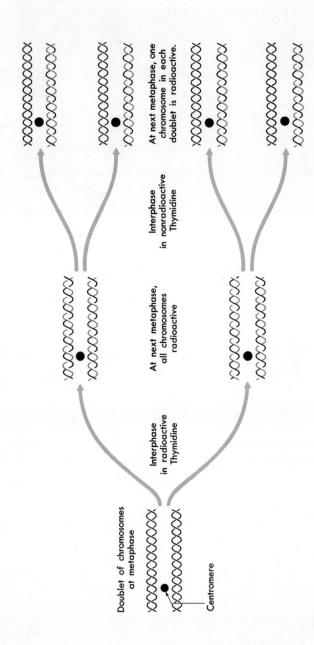

Fig. 7-8

The duplication of chromosomes. The results suggest that (1) each chromosome (chromatid) consists of *two* parts and (2) each of these two parts remains intact during the process of duplication. The parallel between this experiment and the one in Fig. 7-4 suggests that the two parts are the two strands of a DNA molecule and that they separate during interphase according to the Watson-Crick model. Exactly how the DNA is organized in the chromosome is not yet known. This experiment was first performed by Dr. J. Herbert Taylor of Columbia University.

active, the other not (Fig. 7–8). Thus the phenomenon established by Meselson and Stahl in the duplication of DNA molecules also occurs in the duplication of chromosomes.

7–5 SEXUAL REPRODUCTION

In the cells of most higher organisms, each chromosome is represented twice. Such pairs of chromosomes are known as *homologous pairs* or *homologues*. Their origin is quite clear: one homologue was received from the organism's father, the other from its mother. This occurred as the sex cells of the two parents fused in the process of sexual reproduction.

Each species has a characteristic number of chromosomes in its cells. The onion, for example, has 16 (as 8 homologous pairs); the mouse has 40. Man has 46 (Fig. 7–9). In order that this number (called the *diploid* number) be preserved, the sex cells of the organism must have only one-half of this number. Thus the sperm and egg of humans must each have 23 chromosomes if, after fertilization, the offspring is to have its normal complement of 46.

The process by which the sex cells receive the reduced (*haploid*) number of chromosomes is called **meiosis** (Fig. 7–10). It consists of two consecutive cell divisions with but one duplication of the chromosomes. In the first division, there is no duplication of the centromeres. Instead, the duplicated chromosomes (which make up a "doublet") remain together as the centromere moves to the pole of the cell. However, there *is* a separation of the homologous pairs. One homologue travels to one pole while the other goes to the opposite pole. The second division is simply a mitotic division. The centromeres duplicate. Then, as in mitosis, the daughter centromeres move apart and pull their attached chromosome to their respective pole. Meiosis thus produces cells with one-half the normal number of chromosomes. By this reduction in the number of chromosomes, the stage is set for the union of two gametes. Meiosis, then, provides a mechanism by which the traits of *two different parents* can be combined. Whereas cell division by mitosis produces daughter cells identical to the parent cell, meiosis followed by fertilization permits the formation of cells containing an altered hereditary blueprint.

Furthermore, the cells that any one parent produces by meiosis are themselves not alike. Each cell produced by meiosis contains a randomly shuffled assortment of the chromosomes that were present (in pairs) in the parent cell. In Fig. 7–10, for the sake of simplicity, only one pair of homologues is shown. In most living things, the diploid number is larger. The parasitic roundworm *Ascaris bivalens* has a diploid number of 4. These four chromosomes make up two homologous pairs. One member of each pair came from the worm's father, the other from its mother. Thus we can indicate the two pairs as $A^m A^f$ and $B^m B^f$. When *Ascaris* cells undergo the first meiotic division, the homologous pairs orient themselves on either side of the "equator" of the cell. If A^m and B^m orient on one side and A^f and B^f on the other, the gametes will have the same chromo-

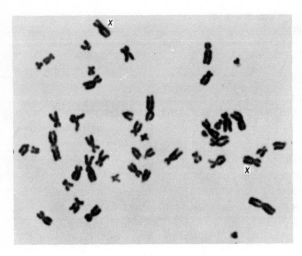

(a)

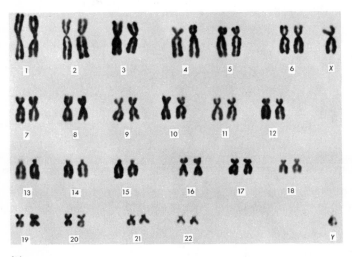

(b)

Fig. 7–9

(a) The 46 chromosomes of a human female. (Courtesy Dr. T. T. Puck and J. H. Tjio. A *karyotype* (b) is prepared by cutting the photograph into pieces containing single chromosomes and arranging them by homologous pairs. This karyotype is of a normal male. (Courtesy Dr. James L. German, III.)

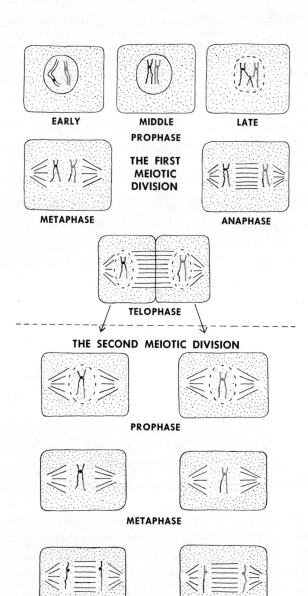

EARLY   MIDDLE   LATE

PROPHASE

THE FIRST
MEIOTIC
DIVISION

METAPHASE   ANAPHASE

TELOPHASE

THE SECOND MEIOTIC DIVISION

PROPHASE

METAPHASE

ANAPHASE

TELOPHASE

Fig. 7–10

Meiosis. The behavior of just a
single pair of homologous
chromosomes is shown with one
homologue black, the other in
color.

some combinations as those the worm received from its parents. It is just as likely, however, that A^m and B^f will orient on one side and A^f and B^m on the other (Fig. 7-11). The gametes produced by this arrangement will contain a new combination of chromosomes. Any organism whose diploid number is 4 can thus produce four kinds of gametes: A^mB^m, A^fB^f, A^mB^f, or A^fB^m. The fly *Drosophila melanogaster* has eight chromosomes in its diploid cells, four derived from its mother and four derived from its father. A little figuring with pencil and paper will show that in distributing one homologue of each pair to the gametes, 16 different combinations can be produced. In fact, the number of combinations produced by this process of *random assortment* is equal to 2^n, where n is the haploid number of chromosomes in the organism. Human mothers and fathers, each with a haploid number of 23, can each produce by random assortment 2^{23} or 8,388,608 different kinds of gametes.

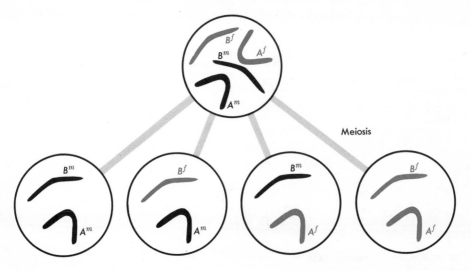

Fig. 7-11

Random assortment of maternal (black) and paternal (color) chromosomes during meiosis in *Ascaris bivalens* ($2n = 4$).

And this is only part of the story. Meiosis provides another mechanism by which the hereditary blueprints of the organism are thoroughly reshuffled. Early in the first meiotic division, the homologous doublets come together in the cell. As they lie united lengthwise, two of the arms, one from each doublet, "cross over." In so doing, they exchange one or two sections. Consequently, only two of the four cells produced by meiosis will contain a "pure" maternal or paternal homologue of any given pair (Fig. 7-10).

7–6 PARALLEL BEHAVIOR OF GENES AND CHROMOSOMES

While it is clear that meiosis reshuffles chromosome segments as well as entire chromosomes, what evidence is there that hereditary traits are reshuffled at the same time? Even before the discovery that the chromosomes of higher organisms occur in pairs, it was known that the hereditary determinants—genes—occur in pairs. And, as with chromosomes, one member of each pair is inherited from each parent. A strain of corn (maize) can be developed that *invariably* produces yellow kernels which are well filled with a food reserve tissue called endosperm. The constancy of these traits suggests that the two genes controlling each of them are alike: *CC*, for yellow color, and *SS* for smooth texture. A second strain can be developed that is pure-breeding for colorless kernels (*cc*) and shrunken endosperm (*ss*). During meiosis the members of these pairs become separated. The gametes of the first strain thus contain one gene for color and one gene for texture (*CS*). The gametes of the second strain have *cs*. Union of these gametes by cross-breeding the two strains results in a plant with *CcSs*. Interestingly enough, the kernels produced by such a plant are both yellow and smooth. We say that the genes for these manifestations of kernel color and texture are *dominant*. The genes for colorlessness and shrunken endosperm are said to be *recessive*.

	Hybrid gametes		
CS	cs	Cs	cS
cs	CcSs	ccss	Ccss	ccSs
Appearance	yellow, smooth	colorless, shrunken	yellow, shrunken	colorless, smooth
If random assortment	25%	25%	25%	25%
If complete linkage	50%	50%	0	0
Actual results	48.2%	48.2%	1.8%	1.8%

Fig. 7–12
Linkage in corn.

Now if this plant, called a hybrid, is crossed with a pure-breeding double recessive (*ccss*) plant, a number of possibilities might be expected. If the genes for kernel color are located on one pair of the corn plant's 10 pairs of homologues and the genes for kernel texture on a different pair, then, because of the random assortment of chromosomes during meiosis, we would expect the hybrid plant to produce four kinds of gametes and in equal numbers (Fig. 7–11). Union of these gametes with the gametes of the *ccss* plant (all of these being *cs*), would produce four different kinds of plants and in equal numbers (Fig. 7–12). Two of the kinds would resemble the parents and two, the *recombinants*, would show new combinations of the traits.

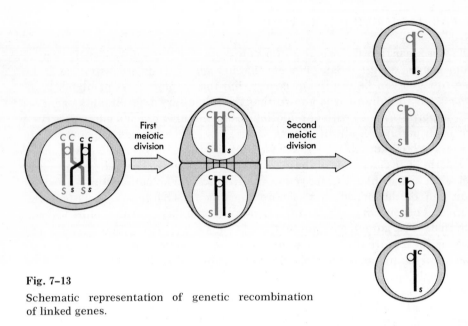

Fig. 7–13

Schematic representation of genetic recombination of linked genes.

On the other hand, perhaps these two aspects of the corn kernel are insep-arable. Perhaps the same genes that produce yellow color also produce smooth texture and the genes for colorlessness also produce shrunken endosperm. In other words, what if only a single pair of genes were involved? Then only two kinds of gametes could be produced by the hybrid plant and only two kinds of offspring would be produced, the parental types (Fig. 7–12).

Actually, neither of these results occurs when such a mating is performed. While most of the offspring are of the parental types (Fig. 7–12), some recombi-nants *are* formed. The strong tendency for the genes to remain associated can be explained by assuming that the genes controlling kernel color are on the same chromosome as the genes controlling kernel texture. Such genes are said to be *linked*. But what of the recombinants? Their case could be explained if the crossing over that occurs in the first meiotic division occasionally separated the genes (Fig. 7–13). If a crossover should occur between the gene for kernel color and the gene for kernel texture, the original combination (*CS* and *cs*) would be broken up and a chromosome containing *Cs* and one containing *cS* produced.

7–7 THE EVIDENCE OF CREIGHTON AND McCLINTOCK

A clear demonstration that the recombination of linked genes occurs as a result of crossing over came in 1932 from the elegant work of the geneticists Harriet Creighton and Barbara McClintock. During the course of their studies of linkage in corn, they developed a strain of corn that had one chromosome (out of 10

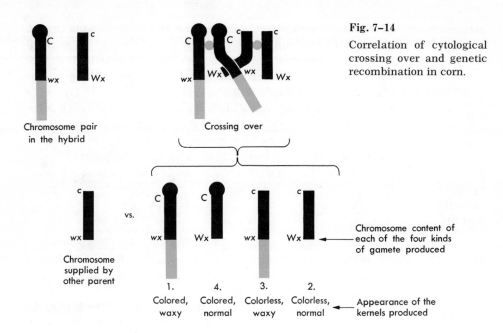

Fig. 7-14

Correlation of cytological crossing over and genetic recombination in corn.

pairs) with two unusual features: a knob at one end of the chromosome and an extra piece of chromosome on the other. As luck would have it, this unusual chromosome carried the gene for colored kernels (C) as well as another gene, a recessive, for waxy endosperm (wx). Its homologue, which was perfectly normal in appearance, carried the recessive gene for colorless kernels (c) and the dominant gene for normal endosperm (Wx). Thus the plant was hybrid for both of these linked traits. In addition, the homologues carrying these genes could be distinguished by microscopic examination because one of them was visibly marked at each end.

These workers reasoned that this plant would produce four kinds of gametes: the parental kinds (Cwx and cWx) and the recombinant kinds (cwx and CWx) produced by crossing over. Fertilization by gametes containing a chromosome of normal appearance and both recessive genes (c and wx) should produce four kinds of kernels: (1) colored waxy (Cc wx wx) kernels, (2) colorless kernels with normal endosperm (cc Wx wx), and the recombinant types: (3) colorless waxy (cc wx wx) and (4) colored kernels with normal endosperm (Cc Wx wx). Furthermore, microscopic examination of the cells of each of the plants growing from these four kinds of kernels should reveal the following kinds of chromosomes. In the first case, there should be one normal chromosome and one extra-long chromosome with the knob at the end. In the second case, both chromosomes should be of normal appearance. However, in the third case, where gene recombination had occurred, one would hope to find evidence that a physical exchange of parts between the homologous chromosomes of the parent had also occurred. Either a normal-length chromosome with a knob at one end should be present, or an extra-long chromosome with no knob. Creighton and McClintock found the latter, thus indicating that the location of the gene (its *locus*) for *wx* was near the end of the chromosome to which the extra segment was attached. The gene locus for kernel color must then be nearer the end with the knob. Examination of the plants in Class 4 (colored kernels, *C,* and normal endosperm, *Wx*) revealed a chromosome of normal length but with a knob at one end (Fig. 7–14). Thus the behavior of the genes, as revealed by examining the inheritance of these traits, was shown to be directly related to the behavior of a particular pair of chromosomes, as revealed by microscopic examination. The recombination of linked genes was shown to occur at the same time that a pair of homologues exchanged parts.

7–8 CHROMOSOME MAPS

The percentage of recombinant gametes in Creighton and McClintock's experiment was about 33%. In the case of the loci *C* and *S,* you remember, the figure was about 3%. Assuming for a moment that the gene loci are in a linear order from one end of a chromosome to the other, we may deduce that the higher the percentage of recombinant gametes formed for a given pair of traits, the greater the distance separating the two loci. The opposite is also true. Although we know that the loci *C* and *Wx* are farther apart than the loci *C* and *S,* we do not yet know whether the locus *Wx* is on the same side of *C* as *S* or on the opposite side (Fig. 7–15). The answer can be determined by developing a corn plant that is hybrid for kernel texture (*Ss*) and kind of endosperm (*Wx wx*). If the percentage of recombinant gametes produced by this individual is less than 33%, then the gene locus *Wx* must be on the same side of locus *C* as locus *S.* The opposite is true if the percentage of recombinant gametes turns out to be greater than 33%. Actually the number of recombinant gametes formed is about 30%. Thus we know that the sequence of gene loci on this chromosome is *C-S-Wx.* Furthermore, the fact that the sum of the percents of recombinants between *C* and *S* and between *S* and *Wx* is so close to the percent between *C* and *Wx* lends strong support to the idea that the gene loci are arranged in a line along the length of the chromosome (Fig. 7–15). A straight line is the only geometric arrangement in which this simple numerical relationship can exist.

Fig. 7–15

Plotting a linkage map. The production of 30% recombinant gametes as a result of crossing over between the *S* and *wx* loci tells us that locus *wx* is on the same side of locus *C* as is locus *S.*

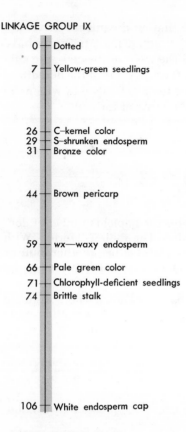

LINKAGE GROUP IX

0 — Dotted

7 — Yellow-green seedlings

26 — C–kernel color
29 — S–shrunken endosperm
31 — Bronze color

44 — Brown pericarp

59 — wx—waxy endosperm

66 — Pale green color
71 — Chlorophyll-deficient seedlings
74 — Brittle stalk

106 — White endosperm cap

Fig. 7–16

Map of one of the chromosomes found in the corn plant (*Zea mays*).

By pursuing this method with as many linked genes on a given chromosome as can be discovered, it is possible to plot chromosome maps. These maps show the sequence in which the gene loci occur and the relative spacing between them. Such maps have been produced for the chromosomes of a number of experimental organisms. Figure 7–16 shows a map of the known gene loci on the corn chromosome that we have been studying.

One additional point deserves mention. As geneticists have established which genes in a particular organism show linkage and which do not, they have been able to assign each gene to a *linkage group*. All the genes in a given group show linkage with respect to one another. They do not show linkage to genes in other groups. The interesting thing about this is that the *number* of linkage groups in an organism is equal to the number of homologous pairs of chromosomes in that organism. Thus *Drosophila melanogaster,* with its four pairs of chromosomes, has four linkage groups. The corn plant, as we have seen, has 10 pairs of chromosomes. All of its *genes* fall into one or another of 10 linkage groups. Similarly, in *Neurospora* and the tomato plant, both of which have been exhaustively studied by geneticists, the number of linkage groups is equal to the number of homologous pairs of chromosomes—seven in *Neurospora* and 12 in the tomato. Here, then, is powerful evidence that an organism's genes are located in or on the chromosomes in the cells of that organism. And, as we have seen, it is also in the chromosomes that we find the DNA of the cell—the substance that Avery and his co-workers showed was the repository of the hereditary information. In the next chapter, we will examine the mechanism by which the information encoded in DNA is translated into the traits of the cell.

EXERCISES AND PROBLEMS

1   The haploid number of horse chromosomes plus the haploid number of donkey chromosomes can produce a healthy mule. The mule is sterile. With rare exceptions (see Question 2) it cannot breed with either another mule or a horse or a donkey. Can you explain why, in terms of the behavior of the chromosomes during meiosis?

2   Although mules are generally sterile, a few cases are known where a female mule has given birth to a horse (after mating with a horse) or another mule (after mating with a donkey). Using your knowledge of meiosis, can you think of an explanation for these rare events?

3   If a hermaphroditic organism (an organism that has both male and female reproductive organs) fertilizes its own eggs, must all its offspring be identical? Explain.

4   How many kinds of gametes can the onion ($2n = 16$) produce by random assortment alone?

5   The gene for tallness in peas is dominant over the gene for dwarfism. A cross between a tall pea and a dwarf pea produced 86 tall plants and 81 dwarf plants. What was the probable gene content of the tall plant?

6   When tall tomatoes with red fruit are crossed with dwarf tomatoes with yellow fruit, all the offspring are tall-red. When these plants are mated, only the two original types are found in the next generation. What do we conclude about the genetic control of these two traits?

7   When each of the following hybrids is crossed with the corresponding double recessive, recombinants are produced in the percentage indicated.

$AaBb$	19%	$BbYy$	7%
$XxYy$	10%	$AaXx$	22%

From these data determine the linear sequence and relative spacing of each gene locus on the chromosome.

8   A normal corn plant is crossed with a corn plant having waxy endosperm and a brittle stalk (see Fig. 7–16). All the offspring are normal in appearance. However, when *these* plants are crossed with plants having waxy endosperm and brittle stalks, what kinds of offspring will be produced? What will be the approximate percentage of each kind?

REFERENCES

1   Hotchkiss, R. D., and Esther Weiss, "Transformed Bacteria," *Scientific American,* Reprint No. 18, November, 1956.   The alteration of bacterial heredity by treatment with DNA is described in detail.

2   Peters, J. A., ed., *Classic Papers in Genetics,* Prentice-Hall, Inc., 1959.   Includes the original papers on the discoveries made by Avery and his co-workers, and Watson and Crick.   Other papers on the nature and action of genes are also included.

3   Hanawalt, P. C., and R. H. Haynes, "The Repair of DNA," *Scientific American,* February, 1967.   Describes how bacteria are able to exploit the complementarity of the two strands of DNA in the repair of damage to the molecule.

4   Swanson, C. P., *The Cell,* 2nd ed., Foundations of Modern Biology Series, Prentice-Hall, Inc., 1964.   Mitosis and meiosis are treated in Chapters 5 and 6 respectively.

5   Mazia, D., "How Cells Divide," *Scientific American,* Reprint No. 93, September, 1961.

6   Taylor, J. H., "The Duplication of Chromosomes," *Scientific American,* Reprint No. 60, June, 1958.   Attempts to relate the structure of DNA to the structure of chromosomes.

7   Gabriel, M. L., and S. Fogel, eds., *Great Experiments in Biology,* Prentice-Hall, Inc., 1955.   This paperback contains several original papers on the chromosome theory, including the report by Harriet Creighton and Barbara McClintock of their experiments.

CHAPTER 8

GENE EXPRESSION

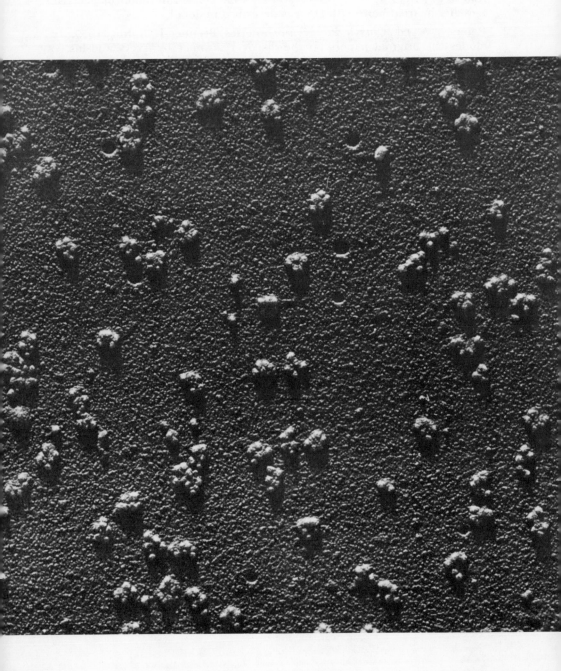

In the last chapter, we determined that genes are made of DNA. However, the polysaccharide capsules of pneumococci and the starchy endosperm in corn kernels are not. These traits are simply the expression or product of gene activity. Let us now examine the mechanism by which genes produce traits.

8-1 THE ONE GENE–ONE ENZYME THEORY

Our knowledge of the method of action of genes grew out of the work of the geneticists G. W. Beadle and E. L. Tatum with the red bread mold *Neurospora*. *Neurospora* is particularly well suited for genetic studies. Its life cycle is shown in Fig. 8-1. Note that throughout most of its life cycle the organism is haploid.

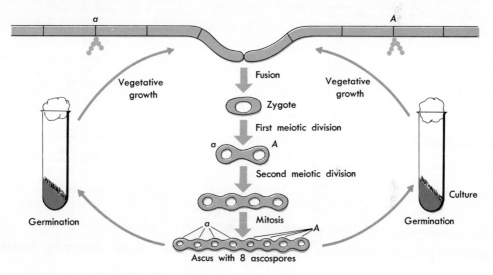

Fig. 8-1

Life cycle of *Neurospora*. The distribution shown here of the *A* and *a* genes would occur only if no exchange of this locus took place during crossing over in the first meiotic division.

Thus each gene is present in only a single dose, and the geneticist does not have to worry about the possibility that recessive genes may be masked by dominant ones. Another virtue is that the meiotic divisions occur in a narrow tube, called an ascus. The tube is sufficiently narrow that the eight nuclei produced by the first and second meiotic divisions, followed by one mitotic division, are not able

◀ Polysomes. They manufacture proteins according to the instructions coded in the genes. (Courtesy Alexander Rich.)

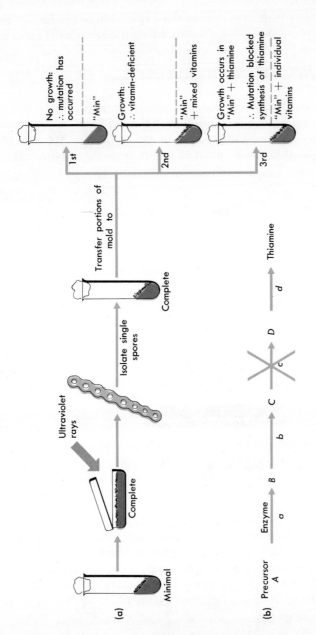

Fig. 8–2

Beadle and Tatum's (a) experiment and (b) hypothesis. Crossing the mutant strain with a normal strain showed that the ability to convert thiamine precursor C into precursor D was controlled by a single gene. Presumably the mutant gene could not produce the necessary enzyme.

to slip past one another. Consequently, if the original diploid nucleus is hybrid for a pair of genes (e.g. *A,a*) and no crossing over at that locus occurs, the genes will segregate at the first meiotic division. After the next two divisions, the ascus will have four spores at one end containing one gene, and four spores at the opposite end containing the other gene. (What patterns might occur if crossing over of the loci *should* occur in the first meiotic division?)

Neurospora can be cultured in a very simple ("minimal") medium. Sucrose, a few salts, and the single vitamin biotin provide all the nutritional requirements for *Neurospora* to live, grow, and reproduce. From these relatively few and simple substances, it is capable of synthesizing all the many complex substances, such as proteins and nucleic acids, that are necessary for life.

Beadle and Tatum exposed *Neurospora* zygotes to ultraviolet rays. Ultraviolet radiation is one of a number of agents that interact with DNA and cause permanent, inheritable alterations in the genetic code. Such alterations are called **mutations.**

The spores produced by these irradiated zygotes were then dissected out individually and each one placed on a "complete" medium, that is, one enriched with many vitamins and amino acids. After some growth had occurred in each culture, portions of the fungus were transferred to tubes of minimal medium (Fig. 8–2). In some cases growth continued; in some cases it did not. When it did not, that particular strain was then supplied with various vitamins, amino acids, etc., until growth did occur. Eventually, each deficient strain was found to be capable of growing on a minimal medium to which *one* accessory substance such as, for example, the vitamin thiamine, had been added. Beadle and Tatum reasoned that the ultraviolet irradiation had caused a mutation, by which a gene that permitted the synthesis of the accessory substance (in this case, thiamine) was transformed into one that did not.

The manufacture of thiamine from the simple substances present in the minimal medium does not take place in a single chemical reaction but in a series of them. Like all chemical reactions in living things, each reaction in the thiamine-synthesis series requires the presence of a specific enzyme. By adding intermediate compounds (precursors) to the medium in which their mold was growing, Beadle and Tatum were able to locate just which step in the synthesis of thiamine was blocked in their mutant strain (Fig. 8–2). If they added to the minimal medium any precursor further along in the process, growth occurred. It did not occur if they added any precursor that preceded the blocked reaction. They reasoned that the change of precursor "C" to precursor "D" was blocked because of the absence of the specific enzyme required. On this basis they created the "one gene–one enzyme" theory of gene action: each gene in an organism controls the production of a specific enzyme. It is these enzymes which then carry out all the metabolic activities of the organism, resulting in the development of a characteristic structure and physiology (e.g. starchy endosperm).

One might argue that Beadle and Tatum's experiment does not prove that only a single gene was involved. However, when the thiamine-deficient mutant is mated with a normal strain, the resulting asci contain four mutant spores and four nonmutant spores. Would this 50:50 ratio occur if more than one gene locus were involved?

The discoveries of Beadle and Tatum have shed new light on a number of human diseases that are known to be hereditary. One of these, alcaptonuria, is a rather rare ailment in which the chief symptom is that the urine of the patient turns black upon exposure to air. Biochemical studies have shown that the disease results when the enzyme that catalyzes the conversion of homogentisic acid to maleylacetoacetic acid (Fig. 8–3) is lacking in the individual. These substances are intermediates in the breakdown of the amino acid phenylalanine into compounds that can be oxidized in the citric acid cycle. When step 4 is blocked, homogentisic acid accumulates in the blood. The kidney excretes this excess in the urine. Oxidation of the homogentisic acid by the air turns the urine black.

Phenylalanine $\xrightarrow{1}$ tyrosine $\xrightarrow{2}$ p-hydroxyphenyl-pyruvate $\xrightarrow{3}$ homogentisic acid $\xrightarrow{4}$ maleylaceto-acetic acid $\xrightarrow{5}$ fumarylacetoacetic acid $\xrightarrow{6}$ fumaric acid + acetoacetic acid

$\downarrow \quad \downarrow$

citric acid cycle

Fig. 8–3

Pathway of phenylalanine metabolism in humans.

Another hereditary disease, phenylketonuria (PKU), is caused by a blockage of step 1. In this case, phenylalanine itself accumulates in the blood. The chief symptoms are a serious stunting of intelligence (most sufferers have to be confined to mental institutions), pale skin, and a tendency to epileptic seizures. The pale skin may result from a lack of tyrosine, from which the pigment melanin (responsible for suntan and freckles) is formed. Both alcaptonuria and phenylketonuria are recessive traits. An individual must inherit a defective gene from both his father and his mother to show the trait. This is quite consistent with the one gene–one enzyme theory: so long as one nonmutant gene is present, the necessary enzyme is manufactured.

Perhaps the most thoroughly analyzed hereditary diseases are those in which abnormal hemoglobin molecules are produced. Hemoglobin is the red, oxygen-carrying pigment found in the red blood corpuscles. Each hemoglobin molecule consists of four polypeptide chains (two "alpha" and two "beta" chains), each with an iron-containing heme group to which the transported oxygen molecule is temporarily attached. Each of the polypeptide chains is made up of about 150 amino acids. The normal hemoglobin of adult humans is known as hemoglobin A (Hb^A). However, almost 100 kinds of abnormal hemoglobins have been discovered in humans. One of the most common of these is hemoglobin S (Hb^S). Individuals who manufacture Hb^S exclusively suffer from a disease called sickle-cell anemia. Their red blood corpuscles become crescent

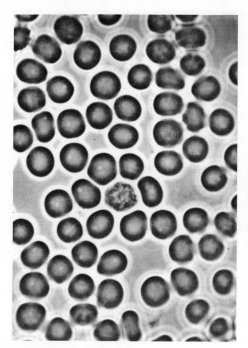 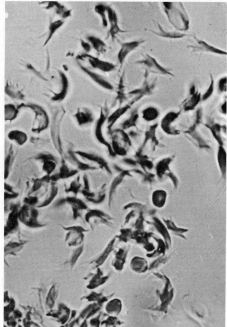

Fig. 8–4

Red blood corpuscles of a victim of sickle-cell anemia. Left: oxygenated. Right: deoxygenated. The shape of the cells when deoxygenated causes them to break easily. (Courtesy Dr. Anthony C. Allison.)

or sickle-shaped, particularly while passing through the capillaries (Fig. 8–4). The distorted corpuscles are very fragile at such a time and are apt to rupture long before their normal life span (about 120 days) is over. Consequently, victims of this disease suffer from a severe and usually fatal anemia.

Victims of sickle-cell anemia inherit a defective gene from each parent. Individuals who inherit only one defective gene have both Hb^S and Hb^A in their red cells. This is not overly disadvantageous. In fact, this combination seems to confer some resistance to one form of malaria which, in turn, probably explains why the mutant gene and the disease are so prevalent in the malarial regions of Africa.

Both Hb^A and Hb^S have been analyzed chemically. It turns out that their alpha chains are identical. So, for that matter, are their beta chains *except* for one "link." At this point, the amino acid valine is present in Hb^S while the amino acid glutamic acid is found at that point in Hb^A (Fig. 8–5).

In almost every one of the other cases of abnormal hemoglobins that have been discovered, the defect occurs as a single amino acid substitution at some point on either the alpha or the beta chain. All the abnormal alpha chains are

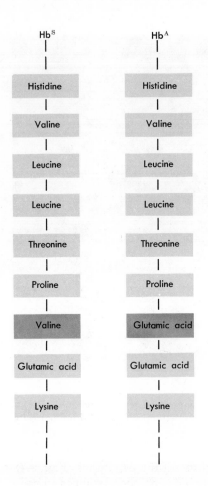

◄ Fig. 8-5

Corresponding portions of the "beta" polypeptide chains of normal hemoglobin (Hb^A) and sickle-cell hemoglobin (Hb^S).

Fig. 8-6 ►

An electron micrograph revealing gene transcription in the developing egg cell of the spotted newt. The long filaments are DNA molecules coated with protein. The fibers extending in clusters from the main axes are molecules of RNA from which the cell's ribosomes will be constructed. Note how transcription begins at one end of each gene, with the RNA molecules getting longer as they proceed toward completion. Note also the large number (up to 100) of RNA molecules that are transcribed simultaneously from each gene. The portions of the DNA bare of RNA appear to be genetically inactive. (Courtesy O. L. Miller, Jr., and Barbara R. Beatty, Biology Division, Oak Ridge National Laboratory.)

the outcome of different mutations within a single gene. The various beta chains represent different mutations in a separate gene. Thus the one gene–one enzyme theory can now be restated in terms of one gene–one polypeptide. This modification also fits the increasing body of evidence that not just enzymes but all the proteins manufactured by cells, e.g. structural proteins like collagen, are the products of gene action.

8-2 PROTEIN SYNTHESIS

Virtually all of the DNA of a cell is located in its nucleus. Most of the protein synthesis in the cell goes on in the cytoplasm. How, then, do the genes control the synthesis of proteins in such a precise way?

In Section 3-6 you learned that protein synthesis occurs at clusters of tiny (150 A) cytoplasmic particles called ribosomes. These are made up of two

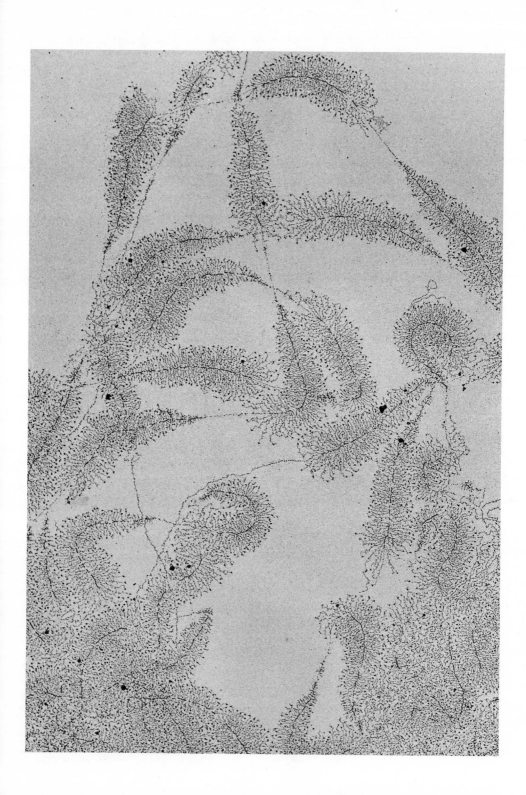

molecules of **ribosomal RNA,** each associated with protein molecules. The RNA molecules, which are synthesized in the nucleolus, are on the surface of the ribosome. There do not appear to be specific ribosomes for the manufacture of specific proteins. Although they are essential to protein synthesis, we must look elsewhere for the instructions by which the nature of the protein being synthesized is determined.

This turns out to be another kind of RNA called **messenger RNA.** Molecules of messenger RNA are synthesized within the nucleus and then pass out into the cytoplasm. They transmit the hereditary message from the DNA within the nucleus to the ribosomes out in the cytoplasm. Thus the name messenger RNA is well chosen.

DNA guides the synthesis of messenger RNA (and ribosomal RNA—see Fig. 8–6) in the same way that it guides its own duplication (see Section 7–3). A single strand of DNA picks up *ribo*nucleotides from the surrounding medium and, with the aid of an enzyme, they are assembled into a strand of RNA whose sequence of bases complements exactly the sequence of bases in the DNA molecule. For every C on the DNA molecule, a G is inserted into the complementary strand of messenger RNA. So, too, every G picks up a C-containing ribonucleotide and every T picks up an A-containing ribonucleotide. The A's on the DNA strand code for the insertion of a uracil-containing ribonucleotide (U). (There is no thymine in RNA.) When the job is completed, the single strand of messenger RNA leaves the DNA as a faithful transcription of it (Fig. 8–7).

This theory of messenger RNA formation has been well supported. When a synthetic DNA consisting solely of T-containing nucleotides ("poly-T") is supplied with all four ribonucleotides (A, U, C, and G), and the necessary linking enzyme, a messenger RNA molecule is formed which contains only adenine in the strand ("poly-A").

Messenger RNA thus transmits the hereditary message in the nucleus to the protein-making machinery of the cytoplasm. Several unanswered questions still remain, however. Proteins are made up of amino acids. How is the sequence of amino acids in a protein related to the sequence of bases on a strand of messenger RNA?

This requires a third kind of RNA called **transfer RNA.** There are perhaps as many as 40–60 distinct kinds of transfer RNA molecules in the cell. Among these there is at least one specific transfer RNA molecule for each of the 20 amino acids used in protein synthesis. With the aid of a specific enzyme and ATP, each amino acid is activated and attached to a particular transfer RNA molecule.

The structure of several transfer RNA molecules has been determined. Each consists of a chain of 75–85 ribonucleotides. Many of these are the normal RNA nucleotides (A, U, G, and C), and in places these bases link two portions of the chain in a double helix like that of the DNA molecule. Because of this, at least one loop is formed in the chain (Fig. 8–7). The ends of the chain are ex-

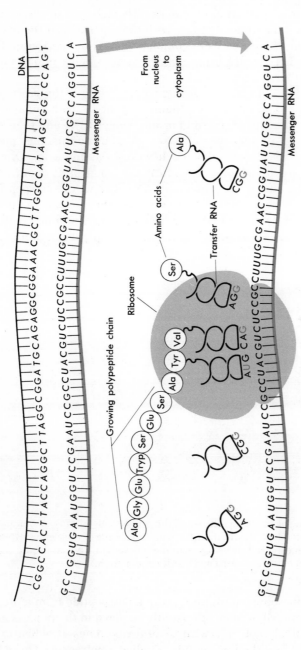

Fig. 8-7

Probable mechanism of the synthesis of a polypeptide according to the genetic instructions coded in DNA. The particular bases with which *transfer* RNA molecules are thought to "recognize" complementary codons on the *messenger* RNA molecule have been tentatively identified for the transfer RNA molecules shown. The use of colored letters for certain bases indicates that the actual base is a chemically modified form of the one indicated but forms base pairs in the same way.

posed, and the amino acid is attached to one of these free ends. It is thought that *unpaired* bases at a loop in the chain "recognize" complementary exposed bases on the messenger RNA molecule and unite with them according to the usual rules of base pairing. Each of 20 kinds of transfer RNA, each kind carrying one specific amino acid, could theoretically unite with those portions of the messenger RNA molecule where the bases were complementary to the exposed bases on the transfer RNA. This would bring specific amino acids into a specific sequence to be joined to form a polypeptide.

The union of transfer RNA and messenger RNA requires the presence of ribosomes. Electron micrographs and biochemical analysis both suggest that the ribosomes attach to the end of a strand of messenger RNA and then move along its length, "reading" its sequence of bases. As they do this, they pick up the appropriate transfer RNA molecules, each with its amino acid. One at a time these amino acids are linked together to form the polypeptide chain. When the ribosome reaches the end of the messenger RNA strand, the polypeptide is complete. It and the ribosome are released. In this way, the message in the DNA molecule—its sequence of bases—which had been *transcribed* into a molecule of messenger RNA becomes *translated* into a specific sequence of amino acids in the polypeptide (Fig. 8–7).

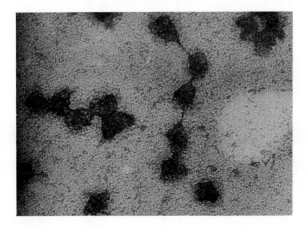

Fig. 8–8

Polysomes from cells engaged in the synthesis of the polypeptide chains of hemoglobin. Five ribosomes are connected by a strand thought to be a molecule of messenger RNA. The length of the polysome suggests that three bases in the messenger RNA molecule code for one amino acid in the polypeptide. (Courtesy Dr. Alexander Rich.)

Although a single ribosome can manufacture a polypeptide from a messenger RNA molecule, several ribosomes are usually engaged in the process at one time. One messenger RNA molecule with several ribosomes attached to it at various stages in assembling the polypeptide is called a **polysome.** These have been observed under the electron microscope (Fig. 8–8).

UUU ⎫ Phenylalanine	CUU ⎫	AUU ⎫ Isoleucine	GUU ⎫
UUC ⎭ (Phe)	CUC ⎪ Leucine	AUC ⎭ (Ileu)	GUC ⎪ Valine
UUA ⎫ Leucine	CUA ⎧ (Leu)	AUA	GUA ⎧ (Val)
UUG ⎭ (Leu)	CUG ⎭	AUG—Methionine (Met)	GUG ⎭
UCU ⎫	CCU ⎫	ACU ⎫	GCU ⎫
UCC ⎪ Serine	CCC ⎪ Proline	ACC ⎪ Threonine	GCC ⎪ Alanine
UCA ⎧ (Ser)	CCA ⎧ (Pro)	ACA ⎧ (Thr)	GCA ⎧ (Ala)
UCG ⎭	CCG ⎭	ACG ⎭	GCG ⎭
UAU ⎫ Tyrosine	CAU ⎫ Histidine	AAU ⎫ Asparagine	GAU ⎫ Aspartic acid
UAC ⎭ (Tyr)	CAC ⎭ (His)	AAC ⎭ (AspN)	GAC ⎭ (Asp)
UAA ⎫ Chain	CAA ⎫ Glutamine	AAA ⎫ Lysine	GAA ⎫ Glutamic acid
UAG ⎭ terminators	CAG ⎭ (GluN)	AAG ⎭ (Lys)	GAG ⎭ (Glu)
UGU ⎫ Cysteine	CGU ⎫	AGU ⎫ Serine	GGU ⎫
UGC ⎭ (Cys)	CGC ⎪ Arginine	AGC ⎭ (Ser)	GGC ⎪ Glycine
UGA—Chain terminator	CGA ⎧ (Arg)	AGA ⎫ Arginine	GGA ⎧ (Gly)
UGG—Tryptophan (Tryp)	CGG ⎭	AGG ⎭ (Arg)	GGG ⎭

Fig. 8-9

The genetic code. Although most of these RNA codons have been assigned as a result of studies with *E. coli,* there is evidence that the code applies to all organisms. In most cases several codons code for a single amino acid, and one codon may be "preferred" by one organism, another by a different organism.

8-3 THE CODE

A great deal of work has also been done in an effort to determine what sequence of bases in the messenger RNA molecule codes each of the amino acids. There are only four kinds of bases (A, U, C, and G) in the messenger RNA molecule, too few for each one to code a single amino acid. Pairs of bases could be arranged in 16 different ways (AA, UU, UA, etc.), but this is still short of the necessary number. Triplets of bases could be arranged in 64 different ways (Fig. 8-9) and thus would provide ample possibilities for coding 20 amino acids. In fact, we might expect two or three different triplets to code a single amino acid. Such a triplet code could also account for the presence of more than 20 different transfer RNA's in a cell with, in some cases, two or three different transfer RNA's each bringing the same amino acid to the growing polypeptide.

By manufacturing synthetic messenger RNA molecules and then supplying them with ribosomes, ATP, enzymes and all 20 amino acids, it has been possible to establish tentatively the triplets that code the various amino acids. For example, a synthetic messenger RNA that has only uracil (poly-U) produces a polypeptide containing the single amino acid phenylalanine. This suggests that the triplet UUU guides the incorporation of phenylalanine into the growing polypeptide chain. Similarly, synthetic poly-A messenger RNA guides the synthesis of a polypeptide containing only the amino acid lysine. Through the use of other synthetic messenger RNA molecules, it has been possible to correlate one or more triplets with each one of the 20 amino acids (Fig. 8-9). Because of this coding relationship, a triplet of bases is called a **codon.**

Three of the 64 possible codons (UAA, UAG, and UGA) have not been found to code for any amino acid. In *E. coli,* at least, these three codons serve as punctuation marks. When the ribosome reaches them, the growth of the polypeptide chain is halted, and the chain is released ready to carry out its function in the cell.

That the genetic code is indeed the simplest possible, that is, a triplet code, has been indirectly supported by the discovery that the polysomes which synthesize the alpha and beta chains of hemoglobin are approximately 1500 A long. The alpha and beta chains each consist of some 150 amino acids. To code these by a triplet code would require 450 bases on the messenger RNA molecule. Since each base occupies about 3.4 A of the chain, this would produce a molecule whose length would be very close (1530 A) to the measured value.

To go back to our one gene–one polypeptide hypothesis, perhaps a mutant gene is simply a section of a DNA molecule in which one base pair has been altered. An alteration of a base pair in the DNA molecule would result in a corresponding alteration in the *messenger* RNA molecule. This, in turn, would provide a codon complementary to a different *transfer* RNA molecule. In some cases this shift might still code for the same amino acid. As you can see in Fig. 8–9, most of the amino acid codons can have the third base of the codon altered without changing the amino acid specified. However, other base changes in the RNA would call for the insertion of a different amino acid into the polypeptide at that point. Such a change is a mutation.

The codons GAA and GAG are thought to code for glutamic acid (Fig. 8–9). In our example of Hb^A and Hb^S, the substitution of a single base, a U for an A in the middle position of the codon, would give a GUA or a GUG codon. Both of these have, indeed, been assigned to valine, the amino acid substitution in sickle-cell hemoglobin. Perhaps, then, the primary difference between a sickle-cell individual and a normal one is the presence of adenine (A) instead of thymine (T) at one spot in the part of a DNA molecule that stores the information for the synthesis of the beta chain of hemoglobin. Although this change seems to be a minor one, the resulting substitution of valine for glutamic acid so alters the physical properties of hemoglobin that a fatal anemia is produced in the individual carrying both genes for the trait.

Not all base changes result in an altered polypeptide; nevertheless, you can well appreciate that even in a "small" gene, such as that specifying the beta chain of hemoglobin, the number of possible mutations is very large. And, in fact, as the study of genes and gene action has progressed, it has become quite clear that a given gene may be mutated in a variety of different ways. In just a few years, almost 100 different mutations have been discovered in the two genes controlling alpha and beta chain synthesis in human hemoglobin. Each one of these results in a polypeptide with a single amino acid substitution somewhere along its length. By means of mutation-producing agents, some 300 different mutations have been produced in a single gene on the chromosome of a virus that infects bacteria.

Is the code universal? Probably so. It appears that the same codons are assigned to the same amino acids in such diverse organisms as *E. coli,* yeast, the tobacco plant, a species of frog, and the guinea pig. However, where two or more alternative codons are available, *which* one is actually used may vary from species to species.

8–4 THE CONTROL OF GENE ACTION

Within its tiny cell, *E. coli* contains all the genetic information it needs to metabolize, grow, and reproduce. As we saw in the last chapter, it can synthesize everything it needs from glucose and a number of inorganic ions. We would expect it to need a large number of enzymes to accomplish so many syntheses, and there may well be 600–800 present in the cell under these conditions. Most of these enzymes, such as the enzymes of cellular respiration, are present at all times. Others, however, are produced only when they are needed by the cell. If, for example, the amino acid arginine is added to the culture, the bacteria soon stop producing the eight enzymes which had been needed to synthesize arginine from intermediates produced in the respiration of glucose. In this case, then, the presence of the *products* of enzyme action *represses* enzyme synthesis.

Conversely, adding new *substrates* to the culture medium may *induce* the formation of new enzymes capable of metabolizing that substrate. If, for example, the disaccharide lactose is substituted for glucose in the culture medium, three new enzymes involved in lactose metabolism begin to appear within a few minutes. One of these enzymes, called a permease, transports the lactose across the cell membrane from the medium into the interior of the cell. A second enzyme, β-galactosidase, hydrolyzes the lactose molecule into the monosaccharides glucose and galactose. Once induced by the presence of lactose, the quantity of β-galactosidase in the cell rises rapidly from virtually none to approximately 3% of the weight of the cell. (The significance of the third enzyme to the metabolism of the cell is not yet known.)

The synthesis of each of these enzymes is controlled by a specific gene. Mapping studies (analogous to those described for corn—see Section 7–8) show that each of these so-called **structural** genes, which clearly are related in function, are also closely linked on the bacterial chromosome. What causes them to begin the work of producing their enzymes? Or rather, what *keeps* them from producing the enzymes before the need arises?

Another gene, called a **regulator** gene, is responsible for this. There is good evidence that the function of the regulator gene is to produce a protein which prevents the structural genes for β-galactosidase, etc., from expressing themselves. Curiously enough, the regulator protein, called the **repressor,** seems not to inactivate the structural genes directly; rather, it represses a gene immediately adjacent to them, called the **operator** (Fig. 8–10). The combination of operator and its associated structural genes is called the **operon.** Perhaps the role of the operator, when it is not repressed by the regulator, is to separate the

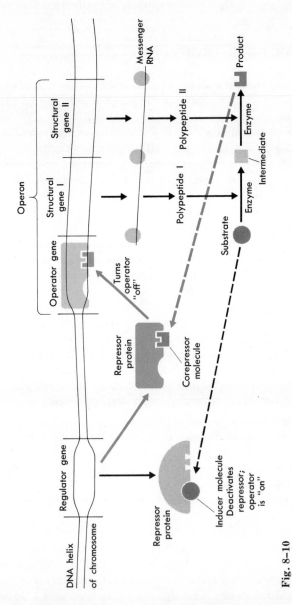

Fig. 8–10

The Jacob-Monod hypothesis of gene control. When an inducer molecule is present, the repressor is unable to attach to the operator gene and polypeptide synthesis takes place. When a corepressor molecule is present, the operon is shut down (color).

two strands of the DNA of the structural genes so that one strand can be transcribed into a single molecule of messenger RNA. Ribosomes moving down this molecule would *translate* the messages into the polypeptides of which the three enzymes are constructed. (You can see why punctuation codons—UAA, UAG, or UGA—would thus be needed to terminate polypeptide synthesis between the portions of the messenger RNA coding for each of the three enzymes.)

What, then, determines whether the repressor substance produced by the regulator gene is active or not? In the case described, it is probable that the presence of lactose itself prevents the repressor from acting on the operator. Lactose may unite with the repressor, a protein, and as a result the shape of the protein may be changed enough so that it can no longer combine with and thus inactivate the operator. The synthesis of β-galactosidase and the other two enzymes may therefore begin.

The mechanism described here was proposed by the French scientists François Jacob and Jacques Monod to explain the genetics of enzyme induction. For this work, they shared in a Nobel prize in 1965.

As we saw above, the action of some genes is repressible. The eight genes involved in the production of the enzymes for arginine synthesis turn out to be clustered in five separate operons. However, a single regulator gene responds to the presence of arginine and represses all five operons. Presumably, the regulator gene produces a repressor which blocks all five operators—when the repressor has combined with arginine, its **corepressor.** The usefulness to the cell of such a mechanism is quite clear. The presence of an essential metabolite turns off the synthesis of the enzymes for its own manufacture and thus stops unnecessary protein synthesis in the cell.

Here, then, is a mechanism by which the genetic code in a bacterium can be *selectively* translated. In these unicellular organisms, the genes are turned off and on in quick response to the changing needs of the cell. What of higher, multicellular organisms? Do similar mechanisms exist in them?

8–5 HORMONES AND GENES

There is a growing body of evidence that similar mechanisms of gene control do occur in the cells of plants and animals. In most of the cases discovered so far, the small molecules that induce or repress gene activity in the cells of these organisms are **hormones.**

During a woman's reproductive years, her body is prepared each month for possible pregnancy. One of the most dramatic changes involved is the preparation of the inner lining of the uterus to receive an embryo. The lining becomes thicker and more richly supplied with blood. The cells that make up the lining enlarge and become very active in protein synthesis. As you might expect, the number of ribosomes in these cells increases markedly. All of these changes are triggered by a rising level of steroid hormones called *estrogens*. There is a growing body of evidence that the primary effect of estrogens on these cells is the derepression of genes. Within two minutes after estrogen molecules reach the cells of the uterus, they enter them, pass through the cytoplasm, and concentrate within the nucleus. At this time, the rate of RNA synthesis in the nucleus begins to increase markedly. New ribosomal and messenger RNA molecules begin to move out into the cytoplasm. Only then does the rate of protein synthesis in the cytoplasm begin to increase.

Another piece of evidence that suggests that the primary effect of estrogens in these cells is on the transcription of DNA is the action of **actinomycin D.** Actinomycin D is an antibiotic that exerts its lethal effect on cells in a remarkably precise way. It unites with intact DNA molecules and prevents the two strands from separating. Thus the DNA molecule cannot serve as the template for the synthesis of either additional DNA (Fig. 7–3) or messenger (Fig. 8–7), transfer, and ribosomal RNA molecules. Without fresh supplies of RNA, especially messenger RNA, the synthesis of new proteins (including all enzymes) by the cell soon comes to a halt. When actinomycin D is administered along with estrogens to the cells of the inner lining of the uterus, all the effects that we have noted fail to occur.

Some hormones may act as gene activators in certain cells, and gene repressors in others. When rat liver cells are treated with the hormone **cortisone,** the synthesis of certain specific enzymes is induced. This induction is preceded by an increase in messenger RNA synthesis. Both of these responses are completely blocked by the simultaneous administration of actinomycin D. This suggests that the primary effect of cortisone in these cells is on the transcription of specific genes for the synthesis of specific proteins. On the other hand, when white blood cells of the rat are treated with cortisone, RNA synthesis is strongly depressed within a minute or two.

A rapidly increasing number of hormones in plants as well as in animals have been shown to produce at least some of their effects through the stimulation or repression of gene activity. Perhaps the Jacob-Monod hypothesis of selective gene action in bacteria applies here as well. Cortisone in liver cells, like metabolic substrates in bacteria, might be an *inducer,* uniting with and deactivating repressor molecules. In this way, operons would be turned "on." Conversely, in a white blood cell, cortisone might be a *corepressor* which, when joined with a repressor protein, blocks gene action.

8-6 THE GENETIC CONTROL OF DIFFERENTIATION

An important problem still remains. A rat, for example, begins life as a fertilized egg. After repeated mitotic divisions, trillions of cells are formed out of which all the structures of the rat are constructed. These cells include not only liver and white blood cells, but nerve, heart muscle, eye lens, skin, connective tissue, secretory cells, and many others. Each of these cells is said to be **differentiated.** Each of the differentiated cell types out of which the organism is constructed has its own characteristic structure and one—or at most a few—specific functions. In the last analysis, these specific structures and functions are probably the result of the synthesis of a few characteristic proteins. Red blood cells manufacture hemoglobin. Heart muscle cells manufacture a contractile protein found nowhere else in the body. The cells of the lens of the eye produce certain proteins, crystallins, which are restricted to these cells and give them their transparency.

As we have seen in the case of the rat, the effect of a given hormone on a cell is determined not simply by the nature of the substance (as it is in a bacterium) but also by the state of differentiation of the cell. But how do cells that have all arisen by mitosis from a common ancestor (the fertilized egg) come to be differentiated in their various ways? One possibility is that during the course of embryonic development, the genes originally present in the fertilized egg become parceled out: liver genes to liver cells, lens genes to lens cells, etc. This is clearly *not* the case. For one thing, both the DNA content and the chromosome count of virtually all the fully differentiated cells of the body are the same— and are the same as in the fertilized egg.

A more direct answer was achieved by the work of Robert Briggs and Thomas King. Using micromanipulators equipped to work on single cells, they succeeded in removing nuclei from the cells of frog embryos in various stages of development. They transplanted these into unfertilized frog eggs, the nucleus of which had been removed. In this way it could be determined whether the nucleus of a cell from a particular stage of embryonic development still retained all the genes of the organism, that is, could guide the development of the egg into a tadpole.

Briggs and King discovered that even after the fertilized egg had developed into a mass of about 16,000 cells, a nucleus from one of these *could* still guide complete embryonic development. However, when the experiments were repeated with nuclei from later embryonic stages, quite different results were obtained. These nuclei were also capable of initiating embryonic development, but with widely varying degrees of eventual success. Many of the resulting embryos ceased developing at one stage or another. Furthermore, those that did develop were often abnormal. This evidence suggests that as embryonic development proceeds, the cell nuclei *do* become altered. Perhaps some of their genes, while not lost, become permanently repressed. However, when nuclei from newly hatched tadpoles are used in these experiments, the results are far more uniform. A few regain their full potentialities and go on to build a complete embryo. This is especially likely to occur with nuclei taken from such rapidly dividing cells as those of the lining of the intestine. What can we conclude from this? It would seem that as differentiation proceeds, no genes are lost, but perhaps certain genes may become permanently incapable of being activated. In this way, the repertory of gene activity that can be evoked in a given differentiated cell becomes severely limited.

How are these decisions made? What determines whether a given cell in the embryo will differentiate into a skin cell or a lens cell? Its location in the

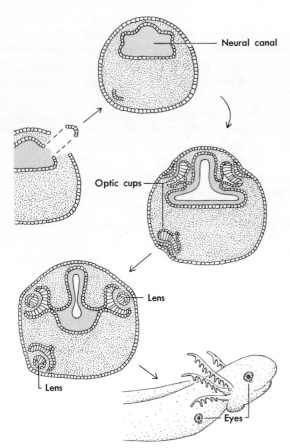

Fig. 8–11

Embryonic development of the salamander eye. A transplanted optic cup induces lens formation in the tissues of its host and an extra eye develops.

embryo appears to be an essential factor. During the development of the brain in salamanders (and other vertebrates) two masses of nervous tissue, the optic cups, grow forward from it. As these near the surface of the embryo, skin cells just in front of them differentiate to form lenses (Fig. 8–11). The optic cups become the retinas of the finished products, the eyes. If a piece of developing brain tissue destined to become an optic cup is transplanted under the flank of a second embryo, an extra eye will form there (Fig. 8–11). We use the term **induction** for the process in which one group of cells thus alters the development of nearby cells.

Induction can occur across a gap filled with agar-agar, suggesting that direct cell contact is not necessary for the process to occur. If, however, a sheet of cellophane is placed between the optic cup and the skin, no induction occurs. This suggests that the process depends upon the movement of macromolecules between the cells.

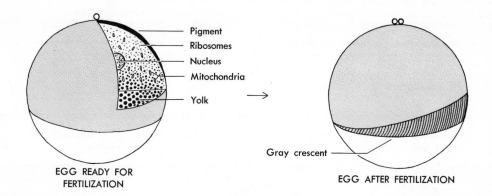

Pigment
Ribosomes
Nucleus
Mitochondria
Yolk

Gray crescent

EGG READY FOR
FERTILIZATION

EGG AFTER FERTILIZATION

Fig. 8–12

Development of the frog egg. Entrance of a sperm cell is followed by the reorganization of cytoplasmic materials into the gray crescent and then the completion of the second meiotic division.

A great deal of work has been done in an attempt to isolate and identify inducing substances, not only in amphibian embryos but in other organisms as well. In some cases, proteins have been implicated. Some of these appear to be ribonucleoproteins, that is, a complex of RNA and protein. In one case, RNA itself is able to induce a new pattern of protein synthesis in the recipient cells.

Although most inducers appear to be macromolecules, there are some exceptions. Vitamin A, a small molecule, can induce altered development in the epidermal cells of a chick embryo. Other small molecules, such as steroid hormones, have also been shown to have inductive properties.

But what induces the first inducer? For one embryonic cell type to liberate its own special product, that cell must already be different from the other cells of the embryo. How can we account for the first appearance of *unlike* daughter cells from the divisions of the fertilized egg?

A closer look at the events occurring in the development of the amphibian egg may help us. The frog or salamander egg is huge as cells go, its volume being about 1.6 million times larger than that of a normal frog cell. The cytoplasm of the egg contains large quantities of DNA, RNA, mitochondria, oil droplets, and yolk. These materials are not distributed uniformly throughout the egg but occur in gradients extending from the dark upper surface of the egg to the white under surface. Shortly after the sperm enters the egg, and *before* the first mitotic division, these cytoplasmic constituents become reorganized. Certain cytoplasmic granules appear at the surface of the egg in a band called the *gray crescent* (Fig. 8–12).

Each mitotic division in the developing egg occurs without growth, so that even after thousands of cells have formed, their total mass is no greater than

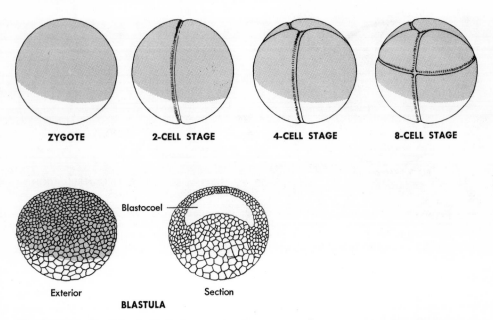

ZYGOTE   2-CELL STAGE   4-CELL STAGE   8-CELL STAGE

Blastocoel

Exterior   Section
BLASTULA

Fig. 8–13
Early stages in the development of a frog.

that of the unfertilized egg (Fig. 8–13). The result of this is that mitosis creates thousands of new nuclei, each partitioned off with a small amount of the cytoplasm of that region of the egg in which it finds itself. Here, then, is a mechanism by which dissimilar cells can arise from a single fertilized egg. And, in fact, the earliest cells known to play an inductive role in the frog are those cells that form in the region of the gray crescent. Something in the cytoplasm of the gray crescent region gives them an identity and special function that sets the stage for the further development of the embryo (Fig. 8–14).

8–7 CHANGING PATTERNS OF GENE ACTION

The importance of the cytoplasm to gene expression has also been demonstrated by studies on developing flies. Some of the large and active cells (for example, in the salivary glands) of these little creatures contain so-called giant chromosomes. These are extended chromosomes existing in hundreds of identical copies and all lying in exact register with one another from end to end (Fig. 8–15). During the course of embryonic development, these chromosomes change their structure. They develop enlarged regions, called "puffs," and it is these regions which are most active in RNA synthesis. Inasmuch as RNA synthesis is the first step in gene action, we may conclude that the enlarged regions of the chromosomes indicate genes or sets of genes that are especially active.

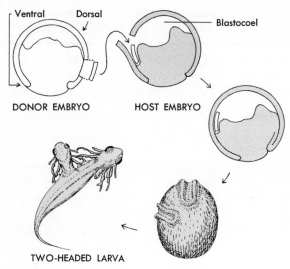

Fig. 8–14

Early induction in the newt embryo. Transplanted tissue (white) from the region of the gray crescent induces the formation of a second head in the host (gray). Most of the tissues of the second head are derived from host cells, *not* transplant cells.

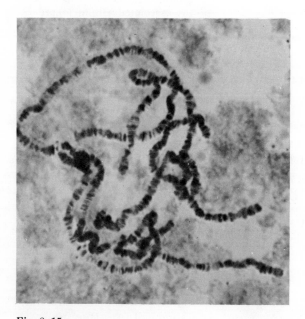

Fig. 8–15

Giant chromosomes from the salivary glands of the fruit fly *Drosophila melanogaster*. Such chromosomes are found in other large, active cells as well. (Courtesy General Biological Supply House, Inc.)

The exact location of the puffs varies over the course of the differentiation of a single kind of cell and also varies from one kind of cell to another. Transplanting a nucleus to another kind of cell or to the same kind of cell at a different stage of development leads to the disappearance of its characteristic puffs and the appearance of new ones. This, then, provides visible evidence of a sequential pattern of gene action during the course of the differentiation of different cell types (Fig. 8–16).

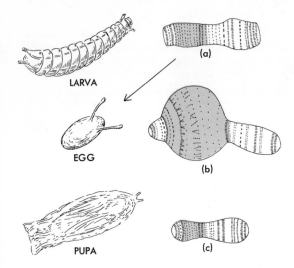

Fig. 8–16

Changes in equivalent portions of a giant chromosome of a fruit fly: (a) in the larva, (c) in the pupa, and (b) when transplanted into an egg. (Eggs do not normally have giant chromosomes.) Enlargements ("puffs") in the chromosomes are associated with increased activity.

In most cases, the nature of the chemical trigger that induces alterations in the pattern of puffs is unknown. One such substance has been discovered, however. It is a steroid hormone, named ecdyson. Because their outer covering is hard, insects can grow only by periodically shedding this covering in the process of molting. This process occurs repeatedly during the period of larval development, which precedes the transformation into the adult. Ecdyson is the substance that triggers each molt. Each time a larva prepares to molt, a definite sequence of puffing occurs in its chromosomes. If ecdyson is administered to a larva shortly after it has completed a molt (a time when its own level of ecdyson is very low), the normal premolt sequence of puffing begins again. It is in these puffs that vigorous RNA synthesis occurs and, in fact, a messenger RNA has been extracted from the nuclei of ecdyson-treated cells that serves as the template for the synthesis of a specific enzyme not found in the cells of untreated larvae.

The production of puffs following the injection of ecdyson is completely blocked by the simultaneous administration of actinomycin D. Here, then, is additional evidence that the hormone acts by unlocking genetic information stored in the cell.

Some RNA synthesis also takes place in the light areas that occur along the giant chromosomes, that is, in the regions between the dark bands (Fig. 8–15).

Here the DNA seems to be in a tenuous, extended state. The dark bands themselves contain a much greater quantity of DNA, but they do not seem to be active in RNA synthesis. Perhaps the DNA in these regions is too tightly coiled and folded to serve as a template for the synthesis of messenger RNA. Perhaps the proteins with which this DNA is united prevent it from being active. You remember that the chromosomes of higher organisms consist of both DNA and protein. Although, as we have seen, the genetic code resides in the DNA, the proteins turn out to have a function as well. These proteins, called **histones,** block the transcription of DNA into RNA. When they are removed from the chromosomes, the rate of RNA synthesis increases markedly. Furthermore, specific proteins not normally made by a given differentiated cell have been shown to be synthesized by the genetic apparatus of that cell when its histones have been removed. This, too, indicates that repressed genes have been derepressed by the removal of histones. The conversion of the undifferentiated cells of the stem bud of a plant into cells that will differentiate to form flowers appears to be triggered by a steroid hormone, the so-called flowering hormone. The first detectable change that occurs in these cells when they are exposed to the hormone is a decrease in their histone content. This is followed by an increase in RNA synthesis which, in turn, is followed by an increase in protein synthesis. Only after these biochemical changes have been initiated do the cells begin to become visibly, i.e. structurally, differentiated.

In general, undifferentiated cells have less of their DNA complexed with histones, and thus kept inactive, than do differentiated cells. In the pea, for example, the amount of active DNA drops from 20% to 10% or less as its cells differentiate. The latter value is about that found in most of the fully differentiated cells of other plants, such as corn, and also of animals such as the chick. In fact, one of the most fully differentiated, and thus specialized, cells of the chick is the red blood cell. It really has only one protein to synthesize—hemoglobin. Virtually 100% of the DNA in the nucleus of a chick red blood cell is complexed with histone and carries on no RNA synthesis. As for mammals, they do not even retain the nucleus once their red blood cells are fully differentiated.

While it is clear that histones block gene transcription in higher organisms, can they alone account for the precision with which specific genes or gene complexes are turned on and off during embryonic development (and after)? Probably not. There is, however, growing evidence that the gene repressors are not histones alone, but histones complexed with RNA. This is a particularly attractive possibility because it easily accounts for the precision of gene repression. The RNA portion of the molecule could carry just the codons needed to "recognize" a given sequence of bases in DNA. There *is* evidence of a fourth kind of RNA (neither ribosomal, nor transfer, nor messenger) that is found attached to chromosomes and perhaps it is part of a ribonucleoprotein repressor.

Not only would a ribonucleoprotein repressor provide a means of recognizing specific genes (through its RNA portion), but its protein part could unite with small molecules (e.g. vitamin A, estrogens, ecdyson, cortisone, the flowering hormone) that have inductive or corepressor properties. Such a repressor molecule would thus resemble in its action the repressors in bacterial cells (Fig. 8–10). Furthermore, it could explain the evidence that macromolecules also play a role in embryonic induction. Perhaps when these substances, proteins or ribonucleoproteins (see Section 8–6), pass from one cell (e.g. in the optic cup) to another (e.g. an epidermal cell), they act as repressors of *regulator* genes, thus turning on operons.

As this discussion has shown, attempts to find out what controls the differentiation of cells have raised more questions than they have answered. Only time will tell how well the Jacob-Monod hypothesis will be able to account for the selective and sequential transcription of the genetic code that occurs during differentiation. Whatever the outcome, though, this hypothesis will have played a crucial role in suggesting new experiments to be performed and posing new questions to be asked. In this way we may approach ever more closely to an intimate understanding of the life of the cell.

EXERCISES AND PROBLEMS

1   Neglecting crossing over, how many different kinds of ascospores can *Neurospora crassa* form by random assortment of its chromosomes? The haploid number is 7.

2   Besides DNA and RNA, what other important biological substances contain the purine adenine?

3   What sequence of bases in messenger RNA will be coded by the following triplets in the DNA molecule: GCT, GAT, CCA, AAA, AGT?

4   Why are haploid organisms especially useful in genetics studies?

5   What inorganic ions would you expect to add to sucrose and biotin to make a minimal medium for growing unmutated *Neurospora*? Why?

6   Using your knowledge of meiosis, predict what arrangements of A spores and a spores could occur in the ascus produced by a *Neurospora* zygote hybrid for these genes.

7   Would you expect mutations in regulator genes to be inherited as dominants or recessives? Explain.

8   Would you expect mutations in operator genes to be inherited as dominants or recessives? Explain.

9   Summarize the mechanisms by which the beef protein you eat is converted into proteins characteristic of you.

REFERENCES

1   Beadle, G. W., "The Genes of Men and Molds," *Scientific American,* Reprint No. 1, September, 1948. The techniques of using *Neurospora* as a tool for studying gene action are described in detail.

2   Ingram, V. M., "How Do Genes Act?" *Scientific American,* Reprint No. 104, January, 1958. Describes the evidence from sickle-cell anemia.

3   Nirenberg, M. W., "The Genetic Code: II," *Scientific American,* Reprint No. 153, March, 1963. Explains how synthetic messenger RNA is used to determine the "letters" in the genetic code.

4   Crick, F. H. C., "The Genetic Code: III," *Scientific American,* Reprint No. 1052, October, 1966. The author, one of the codiscoverers of the structure of DNA, summarizes our current knowledge of the nature of the DNA code and how it is translated.

5   Yanofsky, C., "Gene Structure and Protein Structure," *Scientific American,* May, 1967. Demonstrates that the order and spacing of mutations within a single gene is directly correlated with the order and spacing of amino acid substitutions in the polypeptide produced by that gene.

6   Davidson, E. H., "Hormones and Genes," *Scientific American,* Reprint No. 1013, June, 1965. Presents evidence that the primary effect of many hormones is the activation of genes within the target cells.

7   Fischberg, M., and A. W. Blackler, "How Cells Specialize," *Scientific American,* Reprint No. 94, September, 1961. Shows how the sequence of changes occurring during embryonic development can be traced to specialization within the egg itself.

8   Beermann, W., and U. Clever, "Chromosome Puffs," *Scientific American,* Reprint No. 180, April, 1964. Presents evidence that the puffs are sites of gene activity and shows how the hormone ecdyson alters the pattern of puffing.

9   Bonner, J., *The Molecular Biology of Development,* Oxford University Press, New York, 1965. A superb account of Dr. Bonner's studies on the role of histones in embryonic development. He also shows how many of the results of experimental embryology can be interpreted in terms of the Jacob-Monod theory.

THE METRIC SYSTEM
OF MEASUREMENT

LENGTH

Basic unit is the meter (m), which equals 39.37 in.

kilometer (km)	$= 10^3$ m	micron (μ)	$= 10^{-6}$ m
decimeter (dm)	$= 10^{-1}$ m	millimicron (mμ)	$= 10^{-9}$ m
centimeter (cm)	$= 10^{-2}$ m	angstrom (A)	$= 10^{-10}$ m
millimeter (mm)	$= 10^{-3}$ m		

VOLUME

Basic unit is the cubic decimeter (dm³), which equals 1.06 qt. This volume is more commonly called a liter (l). A liter of water at its maximum density weighs *almost* 1 kg. Therefore, 1 ml (1 ml = 10^{-3} l) of water weighs, for all practical purposes, 1 gm. One cubic centimeter (cm³ or cc) is 10^{-3} dm³. Therefore, it is equal to 1 ml and the units ml and cc are used interchangeably.

MASS

Basic unit is the gram (gm).

kilogram (kg)	$= 10^3$ gm = 2.2 lb	milligram (mg)	$= 10^{-3}$ gm
centigram (cg)	$= 10^{-2}$ gm	microgram (μg)	$= 10^{-6}$ gm

TEMPERATURE

Basic unit is the Celsius (formerly known as centigrade) degree, °C. 0°C is the freezing point of water; 100°C is the boiling point of water. To convert from °C to °F (Fahrenheit) or vice versa: °F − 32 = 9/5 °C.

USEFUL EQUIVALENTS

1 in = 2.54 cm	1 U.S. fluid oz = 29.57 ml
1 oz = 28.35 gm	1 U.S. liquid qt = 0.946 l
1 lb = 453.6 gm	

INDEX